Transport and Energy Processes

Held at the 2024 AIChE Annual Meeting

San Diego, California, USA
27-31 October 2024

ISBN: 979-8-3313-1677-8

Printed from e-media with permission by:

Curran Associates, Inc.
57 Morehouse Lane
Red Hook, NY 12571

AIChE
120 Wall Street, FL 23
New York, NY 10005-4020

Phone: (800) 242-4363
Fax: (203) 775-5177

www.aiche.org

Additional copies of this publication are available from:

Curran Associates, Inc.
57 Morehouse Lane
Red Hook, NY 12571 USA
Phone: 845-758-0400
Fax: 845-758-2633
Email: curran@proceedings.com
Web: www.proceedings.com

TABLE OF CONTENTS

ADVANCED ELECTROCHEMICAL ENERGY STORAGE TECHNOLOGIES I

CO2 CAPTURE, UTILIZATION, AND DISPOSAL: KEY TO CLEAN ENERGY PRODUCTION

POSTER SESSION: TRANSPORT AND ENERGY PROCESSES DIVISION

TEP GRADUATE STUDENT AWARD SESSION

THERMAL ENERGY STORAGE I

TRANSPORT AND ENERGY PROCESSES AT ELECTROCHEMICAL INTERFACES I

TRANSPORT IN ADVANCED FUEL CELL AND WATER ELECTROLYZER TECHNOLOGIES I

TRANSPORT IN ADVANCED FUEL CELL AND WATER ELECTROLYZER TECHNOLOGIES II

ADVANCED ELECTROCHEMICAL ENERGY STORAGE TECHNOLOGIES II

TRANSPORT AND ENERGY PROCESSES AT ELECTROCHEMICAL INTERFACES II

THERMAL ENERGY STORAGE II

ADVANCED ELECTROCHEMICAL ENERGY STORAGE TECHNOLOGIES III

TRANSPORT AND ENERGY PROCESSES AT ELECTROCHEMICAL INTERFACES III

298a Development of Rechargeable Lithium-Sulfur Batteries

Ping Liu

Lithium-sulfur (Li-S) battery is a highly desirable technology featuring high energy density and low-cost materials. However, the challenges with the technology are also well documented. In organic electrolytes, the redox of sulfur goes through a soluble polysulfide mechanism, which greatly impacts the cycle life and practical energy density. In the past few years, we have focused on developing sulfurized polyacrylonitrile, SPAN, as a stable alternative to elemental sulfur-based cathodes. The SPAN reaction does not involve the soluble polysulfide mechanism, which delivers exceptionally long life. Electrolyte engineering, however, remains the key to enable stable lithium metal anode while preserving the stability of SPAN. We show that localized high concentration electrolytes are the optimum choice which form stable interphases on both the Li anode and the SPAN cathode.

SPAN's limitation lies in its limited sulfur content (~ 43 wt%) and specific capacity (< 700 mAh/g). To further raise cell energy density, it is necessary to revisit elemental sulfur. In this regard, we focus on batteries with sulfide-based solid electrolytes. In order to overcome the poor conductivity and large volume change of sulfur cathodes, we discovered a new sulfur-iodine material that features a semiconductor level electronic conductivity and a melting point of 65°C. These features have enabled a solid state Li-S battery with high power density that is also healable. Degradation due to the volume change can be reversed by remelting the cathode, thus restoring performance. We will concluded by discussing remaining challenges for Li-S battery technologies and potential pathways to bring them closer to commercial reality.

298b Processing and Characterization of Thick All Active Material Lithium-Ion Electrodes

Gary Koenig Jr., Chen Cai, Byeongcheol Min

In general, increasing electrode thickness in a battery system will increase energy density at the cell level, because less relative volume is contributed by other inactive but necessary components such as current collectors and separators. While increasing electrode thickness increases volumetric energy density there will be tradeoffs with regards to accessible capacity at increasing rate of charge/discharge and cell overpotential due to the increased ion and electron transport distances for the thicker electrodes. For organic electrolyte systems, such as with lithium-ion batteries, ionic transport limitations are typically more significant, although electronic transport/conduction can also be a factor depending on the specific system and materials involved.

Within our research group, we have developed methods to fabricate thick lithium-ion battery electrodes comprised of only electroactive material. Electrode processing includes thermal treatment to maintain porosity sufficient for percolated ionic transport. While in general the large electrode thickness still results in long ion diffusion paths which limits charge/discharge rate for these electrodes, dramatic reduction of inactive additives provides a relative advantage in effective ionic transport properties. There are also mechanical advantages to processing electrodes with the targeted thicknesses, which can exceed 2 mm.

In this presentation, recent progress for all active material electrode batteries from our group will be discussed. In particular, incorporation of new materials with increased gravimetric and more importantly volumetric energy density will be described. In some cases, the alternative materials have lower electronic conductivity than previously used electroactive materials for all active material electrodes, and processing strategies for alleviating electronic conductivity limitations will be discussed. Incorporation of modified materials and processing results in very high energy density small form factor battery cells, though with significant tradeoffs in terms of rate of charge/discharge.

298c Cathode/Electrolyte Interface in Lithium-Sulfur Batteries Under Lean Electrolyte Condition

Jianjun Chen, Juchen Guo

A low electrolyte/sulfur (E/S) ratio is essential to achieving high specific energy for lithium-sulfur (Li-S) batteries. Comparing to flooded Li-S cells, lean electrolyte condition results in significantly different mass transfer and charge transfer dynamics that closely related to the cathode/electrolyte interfacial properties. We probe the interfacial processes on the sulfur cathode under the lean electrolyte condition using operando electrochemical impedance spectroscopy (EIS) and a galvanostatic intermittent titration technique (GITT). The operando EIS reveals a significant and rapid increase in the charge-transfer resistance during the transition from high-order polysulfides to low-order ones at a low E/S ratio, which is induced by a kinetic bottleneck at the interphase due to Li-ion mass transfer limitation. The GITT results confirm the kinetic bottleneck by revealing a large discharge overpotential during the transition phase. We demonstrate that improving the adsorption of dissolved high-order polysulfides, a key step in the interfacial processes, can alleviate the kinetic limitation, thus enhancing the achievable capacity under the lean electrolyte condition. We further study the effects of ion solvation properties at the interface on the dissolution and reduction of high-order polysulfides.

298d *In-Situ*, Non-Destructive, 3D Neutron Imaging of Li Plating/Stripping at Electrochemical Interfaces in Fast-Charged Li-Ion Batteries

Maha Yusuf, Vanessa Wood, Michael F. Toney, Johanna Weker

Here, we present an *in-situ,* non-destructive 3D characterization of the morphological behavior and spatial heterogeneities of Li plating and stripping on graphite electrodes in full-cell LIBs using high-resolution neutron micro-computed tomography (μ-CT).[4-6] We performed neutron μCT[7] (pixel size: ~ 5.74 μm; effective spatial resolution: ~10-15 μm) at the ICON beamline[8] at the Swiss Spallation Neutron Source at the Paul Scherrer Institute. We imaged two batteries in charged and discharged states: (1) after 4 cycles of 1C charging and (2) after 6 cycles of 6C charging.

Our neutron imaging data shows that Li plates at and around the edge of graphite, indicating that the graphite edge and the areas around the edge are the most susceptible to Li plating. However, certain areas of the cells formed dead Li whereas others formed active Li. Specifically, we examine these spatial heterogeneities in four distinct regions: (1) near Cu current collector (CC), (2) in the middle of graphite electrode, (3) at the graphite-separator interface, and (4) in the separator. Our analysis reveals that Li near the Cu CC remained active, whereas Li near the interface and in the separator became dead at both charging rates: 1C and 6C. In addition, we observed a distinct 3D Li morphology at 6C vs. 1C. Tip-like Li deposits are seen mostly at 6C that became dead following battery discharging, suggesting a correlation between higher XFC-charging rate/cycling number and the increased formation of tip-like dead Li deposits. This work is significant because it deepens our 3D understanding of how Li plating leads to capacity fade, which is needed to realize XFC of LIBs for sustainable transportation.

References:

1. Liu *et al.* (2019). Challenges and opportunities towards fast-charging battery materials. Nature Energy, 1.
2. Liu *et al.* (2016). Understanding undesirable anode lithium plating issues in lithium-ion batteries. RSC advances, 6(91), 88683-88700.
3. Paul *et al.* (2021). A Review of Existing and Emerging Methods for Lithium Detection and Characterization in Li-Ion and Li-Metal Batteries. *Advanced Energy Materials*, *11*(17), 2100372.
4. **Yusuf, M.** (2022). The In-situ Characterization of Fast-charging Degradation Modes in Li-ion Batteries Using High-resolution Neutron Imaging. *The Electrochemical Society Interface*, *31*(4), 38.
5. **Yusuf, M.,** Kaestner, A., Wied, M., LaManna, J. M., Vo, N. T., Dunlop, A. R., ... & Weker, J. N. (2024). Visualizing 3D Morphologies and Spatial Heterogeneities of Li after Fast-Charging via In-situ Neutron Tomography. [Submitted]
6. **Yusuf, M**., Kaestner, A., Zhang, Y., LaManna, J. M., Wied, M., Vo, N., ... & Weker, J. N. (2024, May). (**Energy Technology Division Graduate Student Award sponsored by BioLogic**) In-Situ, Non-Destructive, 3D Neutron Imaging of Lithium Plating Following Extreme Fast-Charging of Full-Cell Lithium-Ion Batteries. In *245th ECS Meeting (May 26-30, 2024)*. ECS.
7. Morgano *et al.* (2018). Unlocking high spatial resolution in neutron imaging through an add-on fibre optics taper. *Optics Express*, *26*(2), 1809-1816.
8. Kaestner *et al.* (2011). The ICON beamline–A facility for cold neutron imaging at SINQ. *Nuclear Instruments and Methods in Physics Research Section A: Accelerators, Spectrometers,*

Detectors and Associated Equipment, *659*(1), 387-393.

298f Impedimetric Sensing of Vented Volatile Organic Compounds before Thermal Runaway of Li-Ion Batteries

Vilas Pol, Sudeshna Bagchi, Amol Bhondekar, Palwinder Kaur

Before the thermal runaway in Li-ion batteries, multiple hazardous and flammable gases including Volatile Organic Compounds (VOCs) are expected to vent. To detect these VOCs/gases quickly and efficiently, Purdue Universities Vilas Pol Energy Research (ViPER) team innovated (1) PEDOT:PSS (poly(3,4-ethylenedioxythiophene) polystyrene sulfonate) based sensor, which is stable up to temperature of 120 degree Celsius. As the PEDOT:PSS based sensor got exposed to various concentrations (5 PPM, 15 PPM, and 30 PPM) of the VOCs, the increase in DC impedance measurement was possible between 1Hz to 1MHz frequency range (2-4).

Principal Component analysis (PCA) was used to classify the responses of the sensor with respect to different analytes or their mixtures (single analyte, binary and ternary mixture) with an accuracy of more than 70%. These preliminary (2-4) research and development activities demonstrated that the initial generation of gases before final thermal runaway of the cells can be tracked in *seconds* mitigating catastrophic event from occurring by *sudden cooling* of the entire system or venting the storage area. The application of such gas sensors coupled with the Battery Management System (BMS) would enhance the safety of Li-ion battery by generating early warning of thermal runaway.

298h Simulation Guided Molecular Design of Hydrofluoroether Solvent for High Energy Batteries

Zhou Yu

Efficient electrolyte design is paramount in advancing next-generation batteries with enhanced energy densities. Hydrofluoroether (HFE) solvents have garnered significant interest due to their potential to offer desirable properties, including high oxidative stability, ionic conductivity, and improved compatibility with lithium metal. Nonetheless, understanding the intricate solvation-property relationships and design principles for high-performance HFE solvents still needs to be improved. This study presents four novel asymmetric HFE designs, systematically varying polyether and fluorocarbon structural building blocks. By leveraging molecular dynamics (MD) modeling to analyze the solvation structures and predict the properties of the corresponding 1 M lithium bis(fluorosulfonyl)imide (LiTFSI) solutions, we downselected the most promising candidate based on solubility, conductivity, and oxidative stability for extensive electrochemical characterizations. The formulated electrolyte exhibits properties consistent with simulation predictions and demonstrates significantly improved capacity retention compared to baseline electrolytes when cycled in lithium metal cells. This research showcases the construction of candidate electrolytes from building block functional moieties to engineer fundamental solvation structures, facilitating the discovery and rational design of novel solvent materials.

104f Thin-Film Composite Membranes for High-Efficiency Nonaqueous Organic Redox Flow Batteries with Enhanced Stability

Taehoon Lee, Zachary Smith

Nonaqueous organic redox flow batteries (NAORFBs) are an emerging class of energy storage systems by featuring high cell voltage, a wide range of operation temperature, and potentially high energy density with low cost. Polymer membrane is a key component to determining efficiencies of NAORFBs by allowing fast transport of supporting ions (*i.e.*, low area-specific resistance, ASR) while preventing crossover of redox species (*i.e.*, low permeance). However, developing a fast and selective membrane has been challenging due to the harsh organic solvent environments which will swell or even dissolve a wide range of polymer structures. Inspired by the industrial success in water purification, herein, we report high-performance thin-film composite (TFC) membranes by *in situ* interfacial polymerization for NAORFBs. Owing to its ultrathin (~200 nm) and crosslinked structure, the prepared TFC membranes exhibited a very low ASR ($<2.3\ \Omega\ cm^2$) coupled with excellent solvent stability compared to those of commercial membranes. Also, a gas chromatography (GC)-based protocol was introduced to screen the small redox species (100–350 g/mol) by evaluating their permeance. Ultimately, we demonstrate the great potential of the developed TFC membranes to boost the efficiency and cycle life of NAORFBs.

47b Integrating Nonaqueous CO2 Capture and Electrochemical Conversion

Reginaldo Gomes, Bidushi Sarkar, Jianping Li, Jiayi Xu, Chukwunwike Iloeje, Cong Liu, Chibueze Amanchukwu

Integrating reactive CO_2 capture and electrochemical conversion offers a promising pathway toward achieving net zero greenhouse gas emissions. However, the effectiveness of amine-CO2 electrochemical reduction in aqueous environments is hindered by the competing hydrogen evolution reaction (HER), primarily due to the abundance of proton sources. Additionally, the generation of carbon-containing products has typically been limited to scenarios involving noble metal catalysts and high CO_2 partial pressures.

In this study, we address these challenges by implementing CO_2 capture and conversion within a nonaqueous medium. Spectroscopic analysis reveals a nearly threefold increase in carbon load compared to aqueous environments. By strategically limiting proton source availability, we achieve an impressive 80% faradaic efficiency (FE) towards CO production using an earth-abundant catalyst. Our techno-economic analysis demonstrates that this enhanced electrochemical performance offsets the higher electrolyte costs associated with nonaqueous media compared to water.

Furthermore, we extend our approach to the reduction of CO_2 found in industrially relevant gas compositions, achieving a notable 40% FE for CO production. Our findings not only contribute valuable insights into electrolyte design for integrated CO_2 capture and conversion but also hold significant implications for addressing CO_2 emissions from industrial sources.

47c Modeling CO_2 reduction and Crossover in Membrane Electrode Assemblies

Eric Lees, Justin Bui, Oyinkansola Romiluyi, Alexis T. Bell, A. Z. Weber

Electrochemical CO_2 reduction is a promising means of using renewable electricity to upgrade waste CO_2 gas into valuable chemicals. To design a CO_2 electrolyzer capable of achieving industrially-relevant product formation rates, one must employ catalyst and membrane materials that enable efficient transport of the reactants (CO_2 and water) and products (CO and hydroxide) to and from the catalyst surface. However, the coupled transport and kinetic processes (*e.g.*, electroosmosis and (bi)carbonate formation) within CO_2 electrolyzers makes selection of these materials challenging because of the tradeoff between product formation rate and utilization of the CO_2 reactant.

This presentation focuses on the development of a 1-dimensional continuum model that is used to resolve CO_2 reduction, (bi)carbonate crossover, and water management in a membrane electrode assembly (MEA) used for converting CO_2 into CO at high current densities. The model is validated using experimentally-measured CO and H_2 formation rates and CO_2 crossover fluxes for an electrolyzer composed of a silver cathode catalyst, anion exchange membrane, iridium anode catalyst, and serpentine flow plates. The model shows how the dominant charge-carrying species in the membrane changes as a function of current density and how the coupling of ion and water transport impacts hydration of the membrane and of the ionomer within the catalyst layers. Simulations reveal the effect of the membrane ion exchange capacity, catalyst layer specific surface area, and electrolyte pH on the CO formation rate and CO_2 utilization in the cell. This talk provides insight into material design strategies for CO_2 electrolyzers and demonstrates the utility of chemical engineering principles for resolving energy losses in electrochemical systems.

47d A Multiphase CFD Model for Electrochemical CO_2 Conversion in a Zero-Gap Membrane-Electrode Assembly Electrolyzer

SeyedSepehr Mostafayi, Ahmad Ijaz, Seyed Parsa Amouzesh, Javad Abbasian, Mohammad Asadi, Hamid Arastoopour

As part of the pathway to net zero carbon economy, the electrochemical conversion of CO2 to different valuable chemicals and fuels using zero-gap membrane-electrode assembly electrolyzers received a significant attention in recent years. However, considerable activity improvement of this technology requires comprehensive studies on the catalyst characterization, cell configuration and operating conditions and flow parameters to achieve effective CO_2 conversion process. The flow behavior and interaction between the liquid and gas phase play crucial role in the performance, selectivity and the stability of the process. Thus, experimentally verified 3-D multiphase model and numerical simulation of CO2 electrolyzers are needed for optimum design, operation and scale up purposes. In this study, the mixture model was used to develop a two-phase steady-state CFD simulation to predict the behavior of gas-liquid flow in the cathode side of a zero-gap CO_2 electrolyzer. The model was also integrated with the chemical and electrochemical equations to predict the current-voltage behavior of the cell, as well as the rate of produced species. The results are validated against the existing experimental data from a lab-scale proof of concept electrolyzer. The developed model can be used to predict the effect of various operating parameters and scale up of this technology.

47e Investigating the Competitive Sorption and Plasticization Effects of Hyperaged 3D Canal Ladder Polymers in Complex CO_2 Gas Mixtures for Carbon Capture Applications

Jing Ying Yeo, Francesco Maria Benedetti, Yan Xia, Zachary Smith

Carbon dioxide (CO_2) separation from natural gas, biogas, and flue gas stands as a critical industrial process. Membrane technology has emerged as an energy-efficient and environmentally benign alternative to traditional amine absorption methods for CO_2 separations. Efforts to improve CO_2 separation have primarily focused on enhancing the affinity of membrane materials for CO_2, often through the incorporation of CO2-affinity functional groups. Herein, we focused on enhancing the CO_2/CH_4 size-sieving effects of microporous hydrocarbon polymers synthesized via catalytic arene-norbornene annulation (CANAL) polymerization, by leveraging on the physical aging process. While typically viewed as an undesired phenomenon in glassy polymers, physical aging here led to a remarkable 3300% boost in CO_2/CH_4 selectivity, with permeability still surpassing that of most commercial membranes even after aging for 180 days. Furthermore, we elucidate substantial enhancements of up to 41% in CO_2/CH_4 selectivity and 50% in combined-acid gas (CAG) selectivity ($CO_2 + H_2S/CH_4$) due to competitive sorption effects observed in mixed-gas tests for a hyperaged 3D CANAL polymer, allowing the materials to surpass the 2018 CO_2/CH_4 mixed-gas and CAG upper bound. Comprehensive evaluations of the stability of the polymeric membrane in the presence of high concentrations of strongly plasticizing CO_2 and H_2S are also conducted in this study. Concentration-dependent CO_2 plasticization reduces CO_2/CH_4 mixed-gas selectivity without affecting CO_2 permeability. Conversely, time-dependent plasticization has minimal impact on CO_2/CH_4 mixed-gas selectivity but increases CO_2 permeability over time. This thorough characterization of these hyperaged CANAL polymers in complex CO_2 mixtures provides invaluable insights into their performance in practical industrial processes, which would otherwise not be captured through typical pure-gas measurements.

47f CFD Model Development of a Small-Scale Fixed-Bed Reactor for Direct Air Capture of CO_2

Hossain Aziz, Bryan Hughes, Subhodeep Banerjee, Mehrdad Shahnam

Direct air capture (DAC) is a promising approach to significantly reduce the CO_2 concentration in the ambient air. This approach requires the extraction of CO_2 from air with ultra dilute CO_2 concentration which makes this approach more expensive than point source CO_2 capture. Several efforts are being made in academics and industry to develop cost-efficient sorbents and solvents suitable for DAC. One of them is polymer with intrinsic microporosity (PIM) which is often used for CO_2 capture because of its higher porosity and rigidity. A novel, cost-efficient aminated PIM sorbent, PF-15-TAEA, was developed at the National Energy Technology Laboratory for DAC applications. The current work is focused on the computational investigation of the CO_2 capture by this novel sorbent from a gas mixture with ultra dilute concentration of CO_2 in dry and humid conditions in a laboratory-scale fixed-bed reactor.

Computational fluid dynamics (CFD) was used to model the adsorption of CO_2 by the PIM sorbent in laboratory-scale fixed-bed reactor. The effects of the presence of solid sorbents in the reactor were incorporated using the porous media model (PMM). The parameters of the PMM were calibrated using the pressure drop across the fixed bed vs. inlet gas velocity data obtained from the experiments. A kinetic model was developed to model the adsorption of CO_2 by solid sorbents from the gas mixture using the experimental data which was included in the CFD model through a user defined function (UDF). The adsorption kinetic model was calibrated by comparing the CO_2 breakthrough curves obtained from experiments and CFD simulations. ANSYS Fluent software was used to perform all the CFD simulations.

Initially, the pressure drop across the bed and the corresponding inlet gas velocity obtained from the experiments was used to present the pressure drop as second-order polynomial (with intercept 0) of velocity which was eventually used to calculate the parameters of the porous media model. This ensured that the pressure drop in the CFD simulations remained in close agreement with the corresponding experimental data. A second-order CO_2 adsorption kinetic model for dry conditions was developed where the equilibrium CO_2 loading was computed by fitting the CO_2 isotherm data obtained from experiments at different temperatures to Langmuir-Freundlich isotherm model. Several CFD simulations were performed with 500 ppm CO_2 in the incoming gas mixture by varying the value of the rate constant of the second-order CO_2 adsorption kinetic model and the CO_2 breakthrough curves obtained from the CFD simulations were compared with the same obtained from the experiments at the same operating conditions. The rate constant value for which the best agreement was obtained between the CO_2 breakthrough curves obtained from experiments and simulations was chosen as the rate constant for second-order CO_2 adsorption kinetic model. This CFD model was used to investigate the effect of inlet gas velocity on the CO_2 adsorption by the aminated PIM sorbents in dry conditions in the fixed-bed reactor.

The PMM-based CFD model was able predict CO_2 adsorption by aminated PIM sorbents from dry gas mixture with 500 ppm CO_2 in a laboratory-scale fixed-bed reactor. Current work is investigating CO_2 adsorption by the aminated PIM sorbent in humid conditions in the laboratory-scale fixed-bed reactor. Our initial experiments show significant improvement in the CO_2 capture performance because of increase in equilibrium CO_2 loading of the aminated PIM sorbent. The CFD model is currently being extended to simulate CO_2 adsorption by the aminated PIM sorbents in humid conditions.

47g Density of CO_2 Loaded Carbon Capture Water Lean Solvents – Measurement and Modeling

Kurt Schmidt, Shu Pan, Joe Costandy, Yiyan Chen, J.W. Farrell

Water lean solvents have recently demonstrated strong advantages for carbon capture compared to traditional aqueous solvents. Robust and accurate process simulation models of the interactions of the solvent and carbon dioxide are critical for equipment sizing, evaluation of different process configurations and operating conditions of carbon capture processes. Obtaining accurate measurements of all the properties at all the relevant conditions of interest for model development is prohibitively time consuming and expensive; as such, a strategy was developed for a pragmatic fluid characterization process. Fluid density is a thermophysical property with significant importance in the process simulation of carbon capture systems and therefore a compelling example to illustrate the fluid characterization process.

The lean and rich solvent density was measured for a water lean solvent with an Anton Paar DMA HPM densitometer over a wide temperature range and carbon dioxide loadings. A novel method was used to prepare the solvent mixtures and the loaded solutions to reduce the uncertainties in the compositions of the fluids being measured. As the solvent is a multicomponent liquid system, the pure and binary mixtures of the constituents of the water lean solvent were also measured to help understand the mixing behavior of the system and to aid in the model development. The individual components of the water lean solvent have limited solubilities in each other, and special attention was made to determine the phase boundaries of the mixtures.

The experimental densities of the pure components, binary mixtures and CO_2 loaded and unloaded solvents were correlated as functions of temperature and composition. The new experimental and model results will not only be useful to evaluate the volumetric properties in water lean solvents systems, but also provide a solid foundation for nearly all properties and process simulation models at relevant operating conditions.

47h CO_2 Hydrogenation to C_{2+} Chemicals on a Metal-Organic Framework (HKUST-1): Mechanistic Investigation

Bikun Zhang, Jianwen Jiang

Thermocatalytic hydrogenation of CO_2 to high value-added multicarbon (C_{2+}) chemicals has received considerable interest to reduce CO_2 footprint and mitigate global warming. With readily tunable structures and functionalities, metal-organic frameworks (MOFs) possess an unprecedentedly large degree of design flexibility and provide an ideal platform to catalyze CO_2 hydrogenation. With Cu paddle-wheel like structure, HKUST-1 has been experimentally reported as a promising catalyst for ethanal production under ambient conditions. Since Cu paddle-wheel is common in MOFs, mechanistic understanding of catalytic mechanism on HKUST-1 would provoke the exploration of more MOFs for C_{2+} chemical production.

In this work, we conduct a theoretical investigation into the mechanism for CO_2 hydrogenation on HKUST-1 through density-functional theory computations. Several atomic models with different structural configurations are considered and a defective HKUST-1 with one 1,3,5-benzene tricarboxylate (BTC) linker removed is identified for facile CO-CHOH coupling via Eley-Rideal mechanism for ethanol production. Intriguingly, only one of the dual Cu atoms is found to serve as an active site for C-C coupling, with the other one as an adsorption site for H atom. Furthermore, the defective HKUST-1 enables coupling of COCHOH and CO intermediates to generate the second C-C bond, which leads to the formation of isopropanol. From bottom-up, this work provides deep microscopic insights into C-C coupling on HKUST-1 and might facilitate the development of new MOFs for efficient CO_2 hydrogenation to C_{2+} chemicals.

574a Biomass Gasification Research for Hydrogen and FT-SAF Production

Jihong Moon, Sang Jun Yoon, Sang Shin Park

SAF (Sustainable Aviation Fuel) production technology includes a variety of renewable raw materials and production technologies, and can be typically divided into HEFA (Hydro-processes Esters and Fatty Acids)-based, alcohol-based, and FT (Fischer-Tropsch)-based SAF. The HEFA method has already been commercialized and is most widely used, but if the bio-jet fuel market grows rapidly by producing liquid fuel through hydrogenation reaction using animal and vegetable fats and oils, bio-jet fuel production technology based on alcohol and FT-SAF technology is needed. In South Korea, plans are being made to demonstrate high-quality syngas production, advanced purification, and supply technology for the FT process that can produce large-capacity next-generation bio-jet fuel using lignocellulosic and herbaceous biomass. Aviation fuel is a very important business in the oil refining industry, and as the electrification of land transportation progresses rapidly, the importance of the next-generation bio-jet fuel market is expected to increase further. The total global biofuel market is expected to more than double by 2050, and demand for eco-friendly fuel is expected to increase significantly in the aviation and shipping sectors. In particular, SAF in the aviation sector will increase from 8 billion liters in 2025 to 4,490 in 2050. Rapid demand growth is expected to reach 100 million liters. In addition, clean hydrogen production technology based on bio-syngas is also being researched. In particular, the production of hydrogen and FT-SAF requires technology to increase hydrogen yield and a high level of impurity removal. The results of experimental research and feasibility analysis will be announced.

574b Effect of Ethanol on the Drying Rate and Rheological Properties of Lithium-Ion Battery Anode Slurries

Min Seo Park, Kyu Hyun

The production process of lithium-ion batteries encompasses various stages, including mixing, coating, drying, and so on. Among these, the drying process is known to be energy-intensive and costly. This is because it requires a fast processing time and is performed at high temperatures to quickly evaporate the solvent.

It is well-known that water is commonly used as a solvent in anode slurry manufacturing. In this study, we focused on the fact that the speed of the drying process can be improved by adding ethanol to water. Improved drying rates reduce energy consumption for solvent removal and improve the economics of anode production. The drying process is a multifaceted process involving not only heat transfer but also mass transfer from materials that make up the slurries, such as particles, binders, and solvents. Within this process, solvent evaporation, binder diffusion, and particle sedimentation occur.

Additionally, ethanol affects the microstructure of the anode slurry. Therefore, in this study, we not only measured the drying rate of the slurry containing ethanol as a solvent, but also investigated the rheological properties to determine the microstructure of the slurry.

To prepare the anode slurries, we used two types of graphite (spherical and flake shape) as active materials, carbon black as a conductive agent, and carboxymethyl cellulose (CMC) as a polymer binder. And we adjusted the water-to-ethanol ratio used as a solvent to 5:5, 6:4, 7:3, 8:2, 9:1, and 10:0.

First, the microstructure of the slurry was examined using an optical microscope (OM) to observe the dispersion of particles. Second, the drying rate of the slurry was measured at room temperature. Finally, steady shear tests and small amplitude oscillation shear (SAOS) tests were performed to investigate the thixotropy, rheological hysteresis properties, and linear viscoelasticity of the slurry, respectively.

This study allowed us to understand how the addition of ethanol as a solvent affects the microstructure, drying rate, and rheological properties of slurries with different active material types. Investigation of the effect of ethanol on anode slurries is important to improve the processability and economics of anode manufacturing.

574c Comparison of Gasification Syngas Composition According to Physical Properties and Type of Biomass

Sang Jun Yoon, Soo Hyun Kim, Tae-Young Mun, Jihong Moon, Sung Jin Park

Biomass gasification is attracting attention as an environmental friendly and efficient utilization technology of renewable biomass energy. In particular, it is recognized as the most effective technology for hydrogen production from solid biomass. Syngas composition produced through biomass gasification is greatly affected by fuel characteristics. In this study, the physicochemical properties of 15 kinds of woody, herbaceous, and organic wastes and syngas composition produced through gasification were measured and compared. Through the analysis of fuel characteristics, all 15 kinds of biomass had a high volatile and oxygen content like general biomass characteristics. Spent coffee ground showed relatively high calorific value because it was partially carbonized through a roasting process. On the other hand, livestock manure showed low calorific value due to high ash and moisture content. Woody and herbaceous biomass showed a calorific value of 17.1 MJ/kg on average. The syngas compositions produced through air gasification of 15 kinds of biomass under reaction temperature condition at 900 °C were compared. In general, the content of H_2 and CO in the syngas produced through the gasification of herbaceous biomass was high. On average, the content of syngas in the product gas was about 60%, and the rest was nitrogen in the air used as a gasification agent. As the volatile content and calorific value of biomass increased, the amount of syngas produced through gasification increased. It was shown that the O/C ratio of the biomass and the amount of syngas produced were inversely proportional.

Acknowledgment

This work was supported by the ERC Center funded by the National Research Foundation of Korea (NRF-2022R1A5A1033719).

574d The Evolution of the Pore Structure-Transport Relationship in Shales Following Microwave Treatment

Agnes Anuka, Mohamed Adam, David Large, Sean P. Rigby

Background

Use of carbon dioxide to displace methane from shale gas reservoirs offers the potential to both enhance gas production but also simultaneously sequester carbon [1]. The process of fracking opens up the previously tight shale pore structure to admit the carbon dioxide to enable displacement of the methane. Microwaves represent a key, efficient technology for fracking the shale rock that offers the potential for better control of the process. However, the exact impact of the microwaves on the pore structure, transport and mineralogy properties of the shale remains uncertain. Further, shales present an ongoing challenge to comprehensive pore structure characterisation due to their complexity [2], as they possess structural and chemical heterogeneities over many length-scales, that all evolve under microwave treatment.

Method

In this work, different shale lithofacies (e.g. Nordland, Utica, and Marcellus) with different ages and levels of thermal maturity, including caprock and reservoir rock types, have been treated with microwaves under a range of conditions (e.g. power, time), and the impact of that treatment on pore structure, transport properties, and mineralogy has been studied. The Nordland, Utica and Marcellus shales studied date from the upper Pliocene, upper Ordovician and middle Devonian, respectively. The Nordland shale is a caprock which has a base ~800 m below sea level. The Utica and Marcellus shales are reservoir rocks.

The pore structural techniques used to study the evolution of the void space are relatively novel, and rarely used for looking at shale and other rocks. The pore structure characterisation has employed the gas overcondensation technique to enable very large macropores to be studied, within the same experiment, in conjunction with the micropores and mesopores typically probed in conventional adsorption experiments [2,3]. This enables the relative juxtaposition and connectivity of pores from a very wide range of size scales to be studied within one technique.

The integrated gas sorption and mercury porosimetry method obtains nitrogen sorption isotherms and rate of adsorption data for both before and after mercury entrapment in a series of experiments conducted on the same sample. This experiment is a means to implement the so-called 'sifting strategy' [2] that attempts to identify the key aspects of the void space that control mass transfer without the need for a 'brute force' approach. The latter requires multi-modal imaging and extensive data management. With the brute force approach, it is often necessary to reduce the raw image data-sets to make mass transport calculations for properly representative volumes tractable with current computing power, often by deriving pore network models. However, it is not possible to know in advance how much information can be dispensed with and still maintain a predictive model. In contrast, the sifting strategy systematically removes different sub-sets of the void space to directly test their variable contribution(s) to mass transport. This can be used to check if the key determining factor for mass transport has been altered due to microwave treatment.

Result

Gas overcondensation, coupled with scanning, has enabled the changes in pore connectivity amongst, and between, the different levels of the porosity to be measured. Simultaneous adsorption and calorimetry experiments have enabled changes in adsorption strength and capacity for carbon dioxide to be correlated with structural changes induced by microwaves, along with the changes to mass uptake properties. Detailed electron microscopy studies have assessed the mineralogical changes, especially mineral grain annealing and illite-smectite clay transformations.

A percolation analysis of the overcondensation data for the Sleipner caprock, Nordland shale, before and after relatively mild (600W, 16s) microwave treatment has been conducted. It was found that microwave treatment leads to there being a greater impact on the mass transfer coefficient, resulting from the removal of a similar percentage of pore volume, consisting of the very largest pore sizes, through mercury entrapment, than for untreated shale. The relative accessibility of the smaller mesopores has been changed by the microwaves, leading to a different impact on structure and mass transport from the loss of larger pores. Studies on more intensively treated Nordland samples (dry, 600 W, 50 s) led to more substantial changes in the void space, with a significant loss of pore volume, as compared to a small growth with lower intensity treatment, and a large relative increase in the general accessibility of the whole void space.

A detailed analysis of the sorption hysteresis obtained via overcondensation experiments has shown the changes to the pore structure in Utica shale. The changes to gas sorption ascending scanning loop data have been used to study changes in the pore network accessibility with microwaving.

Implications

The pore characterization methods used enable changes engendered in the void space by the microwave treatment, not obvious from other techniques, to be quantitatively assessed in a more statistically representative manner than possible with multi-modal imaging. This will enable the changes to producibility of reservoir rocks, or the leakage potential of caprocks, to be better understood.

References:

1.Spanakos, D.; Rigby, S.P. Fuel, 307 (2022) 121928
2. Rigby, S.P. Structural Characterisation of Natural and Industrial Porous Materials: A Manual, Springer International Publishing: Cham, 2020.
3. Rouquerol, F.; Rouquerol, J.; Sing, K. Adsorption by Powders and Porous Solids: Principles, Methodology and Applications; Academic Press: London, 1999.

Acknowledgements:
The authors thank the PTDF, Nigeria, for funding this research.

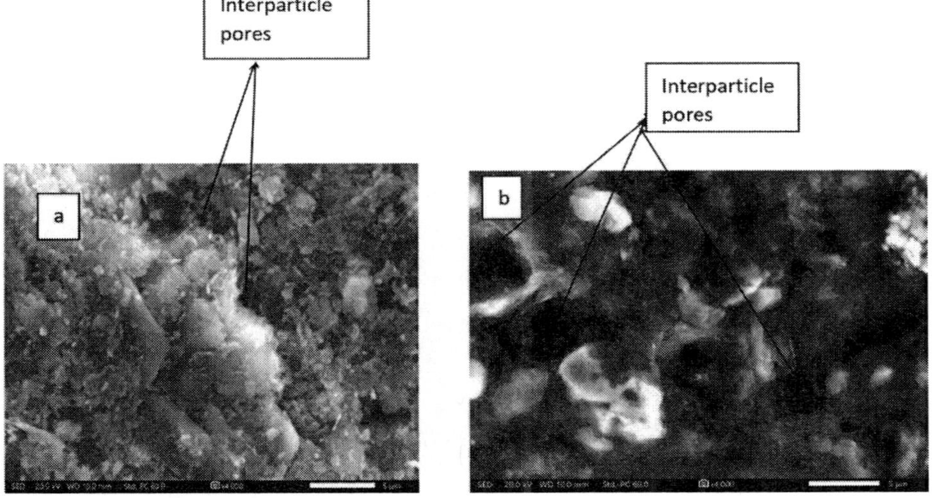

Fig 1. Secondary Electron SEM imaging of (a) Nordland Shale before Microwave Treatment (b) Nordland Shale after Microwave Treatment

574e Molecule Single-File Diffusion through Macroscopic Length of Individual Carbon Nanotube

Jingwei WU

The hydrodynamic behaviors in nanoscale channel have attracted extensive interests in the past decades. Herein, we establish a mass spectroscopy system to investigate the molecule transport in single ultra-long carbon nanotube. Heavy water (D_2O) was step inputted at the entrance of carbon nanotube and the residence time distribution (RTD) was measured by the detector of mass spectrometer. The obtained ion current-time curve was fitted and we find water and gas molecule confined in the nanotube both exhibit ideal plug flow and exit the carbon nanotube at the same time. The surprising result suggests both water and gas molecule form the single-file system and without any back mix. This work not only provides a new method for studying fluid diffusion in the nano-sized domain, but also advances the way to the development of nanofluidics.

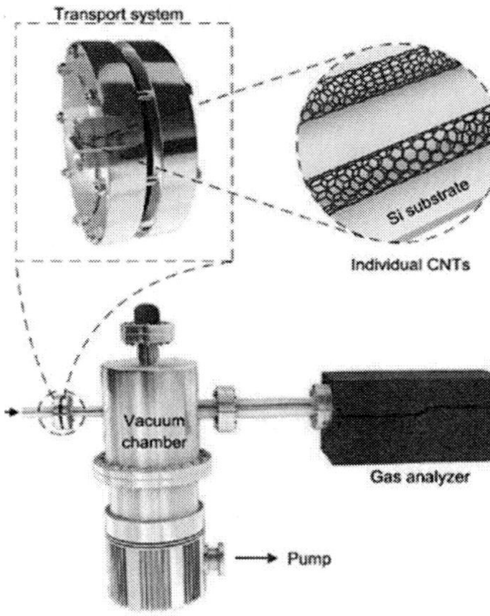

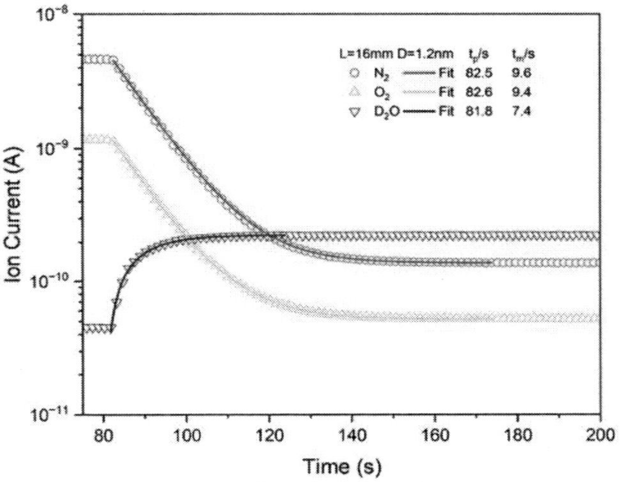

Figure 1: Molecule diffusion through individual carbon nanotube. Ion current vs time curve was fitted by the equation $I=A+B\exp(-(t-t_p)/t_m)$, where t_p and t_m represent the time that fluid molecules pass the carbon nanotube and vacuum chamber respectively.

574f Experimental Characterization of Elastomers for the CO_2 transport Applications

Virginia Signorini, Luca Ansaloni, Thijs Peters, Marco Giacinti Baschetti, Matteo Minelli, Ben Alcock

Carbon Capture and Storage (CCS) technology has gained significant attention following the Paris Agreement and the Climate Change Panel [1,2]. In this concern, CCS technology utilizes either pipelines or ships depending on their economic feasibility and transportation factors such as distances and emission capacity. These different transportation modes require different temperature and pressure conditions, down to cryogenic or supercritical environments, and they have different effects on the polymeric materials [3]. Even though elastomers comprise only a small portion of the CCS value chain since they are mainly used as sealants and gaskets, they are essential for controlling leakage and failure and protecting their mating metal components from corrosive environments [4]. However, polymers may absorb CO_2 molecules that can interact with the matrix by changing the material performances, leading to a plasticizing effect or Rapid gas decompression (RGD) damage [5].

Therefore, this work focuses on proposing a comprehensive approach towards experimental characterization and modeling, specifically tailored to CO_2 transportation in elastomers, with particular attention to Ethylene Propylene Diene Monomer (EPDM), Butyl Rubber (BR), Natural Rubber (NR), and Viton (FKM). The compatibility of elastomers with high-pressure CO_2 is investigated as a function of additives, coupled with different temperatures and pressures for swelling, plasticization phenomena, and mechanical resistance. In fact, CO_2 might lead to an increased backbone mobility and free volume due to the large penetrant solubility, coupled with significant changes in the material mechanical and thermal properties. To this aim, the elastomers have been characterized by thermal, using TGA and DSC, and mechanical DMTA analysis, before and after being exposed to high-pressure CO_2, to better comprehend the correlation between the intrinsic properties of the materials and the strength and durability of the in carbon transport scenarios. In parallel, EPDM, NR, BR, and FKM have been tested via carbon dioxide transient-sorption and direct permeation tests, before and after beign exposed to chloroform for filler extraction, to obtain and compare solubility, diffusivity, and permeability coefficients in mild operating conditions. Large CO_2 uptake can lead to the plasticization of the matrix and, thus, to the reduction of elastomer stiffness, while a large permeability might allow volume deformation that causes geometrical mismatches and seal extrusion [6,7]. Moreover, the research explores the role of reinforcing fillers incorporated into the polymer matrix, evaluating their influence on mechanical, and CO_2 transport properties.

The results demonstrate the extent of polymer plasticization induced by the solubilization of CO_2, at high frequency, which leads to a significant enhancement of the storage and loss energy modulus. Moreover, the higher content of carbon black and additives increases the stress resistance and the stiffness of the polymer, especially for elastomers containing carbon black, such as EPDM and FKM. For what concern transport phenomena, the higher solubility of NR accelerates polymer swelling, further leading to the dilution of low molecular weight compounds or weak intermolecular interactions. On the other hand, the low diffusivity of CO_2 in BR, probably due to the presence of fillers, plays against the wrapped chains, trying to swell the polymer structure which cannot recover after desorption thus leading to blistering phenomena.

These findings provide valuable insights into understanding the performance of elastomers under gas exposure and the complex interaction between supercritical CO_2 and polymers, relevant for the carbon transport chain, in view of their future use in industrial applications.

REFERENCES

1. Erickson, L.E.; Brase, G. Paris Agreement on Climate Change. *Reducing Greenhouse Gas Emissions and Improving Air Quality* **2019**, 11–22, doi:10.1201/9781351116589-2.
2. The Intergovernmental Panel on Climate Change (IPCC) Why the IPCC Was Created.
3. Davies, O.M.; Arnold, J.C.; Sulley, S. The Mechanical Properties of Elastomers in High-Pressure CO 2. **1998**.
4. Nimtz, M.; Klatt, M.; Wiese, B.; Kühn, M.; Joachim Krautz, H. Modelling of the CO2 Process- and Transport Chain in CCS Systems—Examination of Transport and Storage Processes. *Geochemistry* **2010**, *70*, 185–192, doi:10.1016/J.CHEMER.2010.05.011.
5. Schrittesser, B.; Pinter, G.; Schwarz, T.; Kadar, Z.; Nagy, T. Rapid Gas Decompression Performance of Elastomers – A Study of Influencing Testing Parameters. *Procedia Structural Integrity* **2016**, *2*, 1746–1754, doi:10.1016/J.PROSTR.2016.06.220.
6. Swelling of Elastomers in CO2 Environment: Testing Methodology and Experimental Data. | 10.2118/169277-Ms Available online: https://sci-hub.se/https://doi.org/10.2118/169277-MS (accessed on 26 June 2023).
7. Dubois, J.; Grau, E.; Tassaing, T.; Dumon, M. On the CO2 Sorption and Swelling of Elastomers by Supercritical CO2 as Studied by in Situ High Pressure FTIR Microscopy. *J Supercrit Fluids* **2018**, *131*, 150–156, doi:10.1016/J.SUPFLU.2017.09.003.

574g CO2 Impurities Impact on Methanol and Methane Synthesis

Richard Porter, Haroun Mahgerefteh

For the application of carbon capture and utilization in emitting industries (e.g. iron and steel, cement, waste-to-energy, refineries etc.), the impact of CO_2 impurity components within the process chain is a critical issue which must be assessed for successful technology deployment. Flue- and off-gases targeted for capture comprise a range of components other than CO2, including contaminants at low concentrations in the parts per million or parts per billion region (e.g. SO_x, NO_x, Hg etc.). These have the potential to form impurity compounds in resulting CO_2 product streams that are amenable for use as feedstock to CO_2 utilization processes. The range and level of impurities in CO_2 streams captured from different industrial sources present differently depending on industry type, fuels and feedstocks, capture system employed, modes of operation, and additional pollution control systems used.

Selecting an appropriate CO_2 utilization route for a given emission source will also depend on a range of market factors including capacity and product value along with technical factors such as process scale and economics. Important amongst the technical considerations is the presence of impurities in the feedstock CO_2 stream for the conversion process. This is because impurities can have a range of potentially deleterious effects on CO_2 conversion. These effects include i) dilution, mainly from inert gases, ii) modification of catalyst properties, such as by adsorption on, or reaction with, a catalyst surface, iii) competition with CO_2 as reactant, and iv) contamination of products.

In this paper, we discuss how impurities that may exist in captured CO_2 can affect subsequent utilization processes and summarise the progress in this emerging research area. We begin by examining published works that investigate the influence of the interactions between impurity gases and the thermochemical CO_2 utilization processes of methanation and methanol production. These studies include several types of catalysts and involve various concentrations of impurities. We then use the information generated in the comprehensive review to define limits for CO_2 impurities for the CO_2 utilization processes. We next perform analysis using the data generated from the review and impurity limits definitions in order to match the most appropriate capture technologies and sources to specific CO_2 utilisation applications. Suitable technology matches are identified based on the tolerance levels for impurities of specific utilization technologies and estimated cost of CO_2 purification processes.

574h Polarized Composite of SnO2@BaTiO3 Microsphere as an effective Li storage with Enhanced Electrochemical Performance

Hiesang Song

A polarizable composite of BaTiO3-coated SnO2 microsphere (SnO2@BaTiO3) with high porosity was prepared via a solvothermal process and solution coating for application as an anode for high-performance Li-ion batteries. BaTiO3 in the as-formed SnO2@BaTiO3composite generated locally polarized piezoelectric fields under the mechanical stress generated by the volume expansion of SnO2 during the redox process. The synergistic effect of the local piezoelectric potential and high porosity of the SnO2@BaTiO3 composite electrode improved its Li+ mobility as compared to that of the SnO2 microsphere. The improved Li+ mobility of the SnO2@ BaTiO3 composite enhanced its electrochemical performance as the anode for Li-ion batteries.

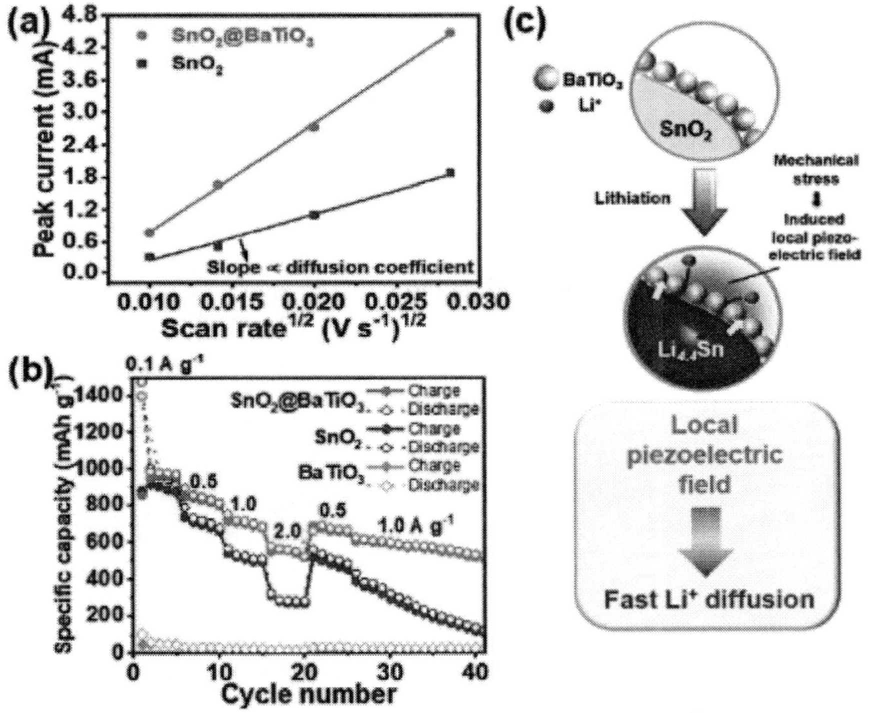

Figure 1. (a) Peak current (I_p) as a function of the square root of the scan rate ($v^{1/2}$) for SnO_2@$BaTiO_3$ composite and SnO_2 microspheres. (b) Rate capability test results at 0.1–2.0 A g^{-1} and subsequent cycling test results at 1.0 A g^{-1}. (c) Schematic diagram of SnO_2@$BaTiO_3$ composite during electrochemical reactions.

104a Ultrafast Deposition of Faceted Lithium Polyhedra By Outpacing SEI Formation

Xintong Yuan, Yuzhang Li

Electrodeposition of lithium (Li) metal is critical for high-energy batteries. However, the simultaneous formation of a surface corrosion film termed the solid electrolyte interphase (SEI) complicates the deposition process, which underpins our poor understanding of Li metal electrodeposition. Here we decouple these two intertwined processes by outpacing SEI formation at ultrafast deposition current densities while also avoiding mass transport limitations. By using cryogenic electron microscopy, we discover the intrinsic deposition morphology of metallic Li to be that of a rhombic dodecahedron, which is surprisingly independent of electrolyte chemistry or current collector substrate (Fig. 1). In a coin cell architecture, these rhombic dodecahedra exhibit near point-contact connectivity with the current collector, which can accelerate inactive Li formation. We propose a pulse-current protocol that overcomes this failure mode by leveraging Li rhombic dodecahedra as nucleation seeds, enabling the subsequent growth of dense Li that improves battery performance compared with a baseline. While Li deposition and SEI formation have always been tightly linked in past studies, our experimental approach enables new opportunities to fundamentally understand these processes decoupled from each other and bring about new insights to engineer better batteries.

Fig. 1. Distinct Li electrodeposition morphology as a function of electrolyte chemistry and current density.

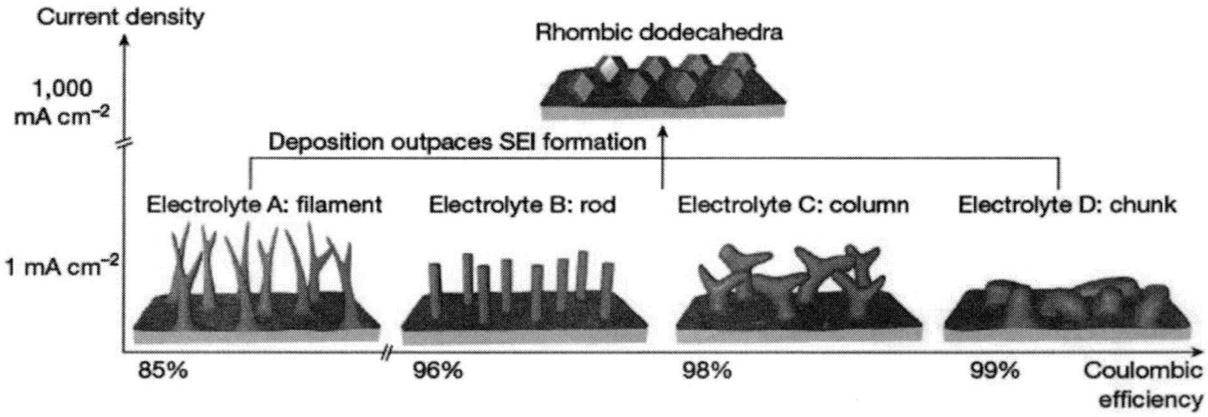

104c Engineering Water Microenvironments in Aprotic Solvent for Enhanced CO2 Electrochemical Reduction

Reginaldo Gomes, Chibueze Amanchukwu

Electrochemical carbon dioxide reduction (CO2R) presents a promising avenue for sustainable production of fuels and chemicals. Water molecules play a crucial role in this process by stabilizing key reaction intermediates and acting as a proton donor. However, the CO2R selectivity is often hindered by the competing hydrogen evolution reaction (HER), even at low water concentrations. In this study, we tune the water microenvironment and electrochemical activity within a series of organic aprotic solvents featuring different functional groups and physicochemical properties.

Through spectroscopic analyses, density functional theory (DFT) calculations, and molecular dynamics (MD) simulations, we establish a direct correlation between solvent donor number and the water hydrogen bonding strength and dynamics. By confining water within a strong hydrogen bond network, we could extend the HER onset potential by nearly 1 V. Using a gold electrode, we could then achieve CO Faradaic efficiency (FE) values of 100% for water concentrations as high as 3 M. Our results showcase the effectiveness of modulating the water microenvironment in enhancing CO2R selectivity. We further extend this approach to mildly acidic conditions, achieving sustained 98% FE towards CO over an earth-abundant zinc catalyst, with virtually no carbonate losses.

By elucidating the interplay between water solvation and electrochemical activity, we offer valuable descriptors to guide efficient electrolyte design across various chemical transformations.

104d Investigating the Performance of Li_2CO_3 in the Li-Metal Solid Electrolyte Interphase Via Gas-Reacted Interphases

Katherine Steinberg, Betar Gallant

Transitioning from graphite to Li-metal anodes would be extremely desirable, offering a roughly 10x increase in the gravimetric energy capacity of Li-ion batteries. However, Li-metal anodes have yet to reach the 99.95+% Coulombic efficiencies (CE) required for long-term stable cycling (80-90% capacity retention for >1000 cycles), because the thermodynamic instability of Li-metal in practical electrolytes leads to parasitic reactivity at the Li surface (1). This reactivity generates the solid electrolyte interphase (SEI), a nanoscale, multicomponent passivation film that governs subsequent behavior of Li interfaces (2). Ideally, the SEI should permit facile conduction of Li^+ ions while inhibiting further electrolyte decomposition, but in practice, electrolyte design for SEI improvement is complicated by the field's lack of data on the conductivity and stability of common SEI materials. Experimental measurements are challenging because the complex nanoscale structure and extremely reductive conditions (-3.04 V vs SHE) of the SEI can dramatically change material properties from bulk behavior (3). In previous work, our group used reactions of Li metal with O_2 or NF_3 to synthesize single-component, nanoscale films of Li_2O or LiF, which enabled direct measurements of transport properties and reactivity of these SEI materials (4-6). Here, we applied similar techniques to study Li_2CO_3. Li_2CO_3 is of particular interest because previous literature has shown CO_2 to be a beneficial additive in certain electrolytes (7), but reductive instability (8) and other possible decomposition pathways (9) call into question whether Li_2CO_3 is capable of effectively passivating Li. In this work, we investigated the functionality of Li_2CO_3 in the SEI using two platforms: synthetic Li_2CO_3-containing films on Li metal, and separately, Li-Cu half cells purged with either Ar or CO_2.

Li_2CO_3-containing films were prepared on Li via sequential reactions of O_2 and CO_2 at slightly elevated temperatures (175-200 °C). The films were confirmed to be conformal, smooth, and relatively pinhole-free using scanning electron microscopy (SEM) and air exposure experiments. A combination of Fourier-transform infrared attenuated total reflectance spectroscopy (FTIR-ATR) and titration-style quantification experiments were used determine the composition of the films. We found that the formation of Li_2CO_3 also generated Li_2O and Li_2C_2, likely due to reduction at the Li interface. The resulting films were stable and passivating in neat solvents and non-fluorinated electrolytes, confirming that Li_2CO_3 and its reduction products are electrically insulating and insoluble. However, the films were found to react in a variety of fluorinated electrolytes, generating materials chemically similar to the native SEI in each electrolyte. Impedance spectroscopy revealed that the Li_2CO_3-containing films were highly ionically conductive, with substantially greater ionic conductivity (~4-12 nS/cm) than previously measured in Li_2O (~1 nS/cm) and LiF (~0.5 nS/cm) (4). To understand the implications of these observations, Li-Cu cells were cycled after either Ar or CO_2 purging in a variety of electrolytes. Across both carbonate- and ether-based electrolytes, fluorinated and nonfluorinated, CO_2 led to substantial improvements in CE for all the electrolytes with the conventional 1 M salt concentration. The exception was local high concentration electrolytes, where CO_2 saturation had no effect. Titration gas chromatography (TGC) was performed to quantify species formed in the resultant SEIs, finding that improvements in CE with CO_2 purging are associated with decreases in accumulation of inactive Li in the SEI. Taken together, these results show that while Li_2CO_3 does exhibit reactivity with fluorinated electrolytes, it can improve the ionic conductivity of Li SEI, leading to more facile Li^+ ion transport and better Li utilization. This work underscores the utility of Li_2CO_3 as an SEI phase and could help inform future electrolyte design for improvement of the Li-metal SEI.

References

1. G. M. Hobold, J. Lopez, R. Guo, N. Minafra, A. Banerjee, Y. Shirley Meng, Y. Shao-Horn and B.

M. Gallant, *Nature Energy*, **6**, 951 (2021).

2. E. Peled and S. Menkin, *Journal of The Electrochemical Society*, **164**, A1703 (2017).

3. R. Guo, G. M. Hobold and B. M. Gallant, *Current Opinion in Solid State and Materials Science*, **26** (2022).

4. R. Guo and B. M. Gallant, *Chemistry of Materials*, **32**, 5525 (2020).

5. R. Guo, D. N. Wang, L. Zuin and B. M. Gallant, *ACS Energy Letters*, **6**, 877 (2021).

6. M. F. He, R. Guo, G. M. Hobold, H. N. Gao and B. M. Gallant, *Proceedings of the National Academy of Sciences of the United States of America*, **117**, 73 (2020).

7. D. Aurbach, A. Zaban, Y. Gofer, E. Ely, I. Weissman, O. Chusid and O. Abramson, *Journal of Power Sources*, **54**, 76 (1995).

8. B. Han, Z. Zhang, Y. C. Zou, K. Xu, G. Y. Xu, H. Wang, H. Meng, Y. H. Deng, J. Li and M. Gu, *Advanced Materials*, **33** (2021).

9. B. S. Parimalam, A. D. MacIntosh, R. Kadam and B. L. Lucht, *Journal of Physical Chemistry C*, **121**, 22733 (2017).

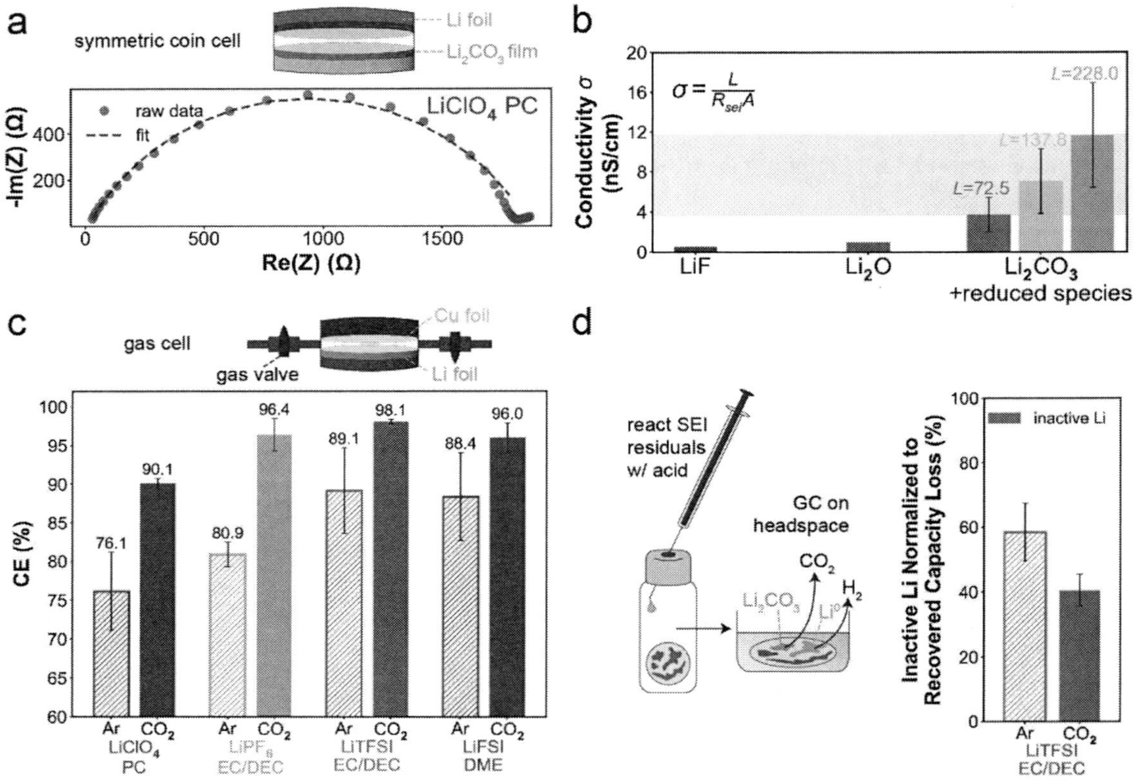

Figure 1. Selected results from investigation of Li_2CO_3 in the Li-metal SEI via gas-reacted interphases. (a) Schematic of symmetric coin cells used to conduct impedance spectroscopy on prepared Li_2CO_3-containing films (top), and Nyquist plot of impedance results (bottom). (b) Ionic conductivity of Li_2CO_3-containing films calculated with a range of film thickness estimates and compared to ionic conductivity results from LiF and Li_2O. (c) Schematic of gas cells used to cycle Li under Ar vs CO_2 saturation (top), and CE results from conducting cycling in a variety of different electrolytes (bottom). (d) Schematic of procedure for TGC quantification of residual SEI materials after cycling (left), and sample TGC inactive Li quantification for 1 M LiTFSI EC/DEC, 1:1 v/v.

104e Stabilizing Cobalt-Free and Ultrahigh-Nickel Cathodes for Lithium-Ion Batteries

Xiaoyang Zhao, Juchen Guo

Abstract:

The rapid ascendancy of lithium-ion batteries (LIBs) as the power source of choice for electric vehicles (EVs) underscores a pressing need for advancements in cathode material technology. Traditional cathodes, often cobalt (Co)-based, are beset with issues like cost volatility, ethical sourcing dilemmas, and environmental detriments, driving the quest for alternative materials. This research, spearheaded by Xiaoyang Zhao from the University of California, Riverside, targets the innovation of cobalt-free and ultrahigh-nickel (Ni) cathodes, aiming to reconcile the economic and environmental facets of battery production with the stringent performance metrics set forth by the U.S. Department of Energy (DOE).

Motivation and Background: The LIB market, integral to the EV industry's growth, is at a critical juncture where the demand for high-energy, cost-effective, and sustainable batteries is at an all-time high. Cobalt, despite its prevalent use in cathode materials, has been problematic due to its price fluctuations, ethical sourcing issues, and environmental concerns. This has necessitated a shift towards alternatives like ultrahigh-Nickel cathodes that promise to achieve DOE's ambitious targets of 235 Wh kg−1 energy density at a cost below $100 per kWh. The adoption of such cathodes could revolutionize energy storage systems, making EVs more competitive with traditional internal combustion engine vehicles.

Methods: The research employed a comprehensive methodology involving the synthesis of ion-mixing-free $LiNiO_2$ (LNO) to study degradation mechanisms and the application of doping and surface modification techniques to enhance cathode stability. The synthesis process was designed to circumvent common pitfalls such as nickel substitutional defects and undesirable phase transitions, using a nuanced two-step approach. Advanced characterization tools like Scanning Electron Microscopy (SEM), Transmission Electron Microscopy (TEM), X-ray Diffraction (XRD), and X-ray Photoelectron Spectroscopy (XPS) were deployed to elucidate the microstructural and electrochemical properties of the cathodes. Electrochemical analysis was conducted to assess the enhancements brought about by doping elements such as Aluminum (Al), Tungsten (W), and Boron (B), in terms of the cathodes' structural stability and electrochemical performance.

Results: The initial phase of the study yielded promising outcomes, with the successful synthesis of ion-mixing-free LNO, indicating a potential reduction in degradation mechanisms. Subsequent doping experiments revealed significant enhancements in the cathodes' performance. Aluminum doping improved rate capability and thermal stability, Tungsten doping alleviated issues related to phase transitions, and Boron doping stabilized the crystal structure. These modifications collectively resulted in cathodes that not only met but exceeded DOE's performance targets, offering improved cycling stability, energy density, and safety. The research illuminated the intricate interplay between material composition, structural integrity, and electrochemical performance, laying the groundwork for future advancements in LIB technology.

Implications: The findings of this study have profound implications for the future of energy storage in EVs and beyond. By developing cobalt-free and ultrahigh-nickel cathodes that align with economic and environmental goals, this research paves the way for the next generation of LIBs. The enhanced understanding of material degradation mechanisms and the effective strategies for cathode stabilization are poised to catalyze further innovations in the field, potentially leading to batteries that are more efficient, cost-effective, and environmentally friendly. As the world gravitates towards sustainable energy solutions, the significance of such advancements in battery technology cannot be overstated, promising a greener, more sustainable future for global transportation and energy storage systems.

In conclusion, this comprehensive study not only addresses the pressing issues associated with cobalt use in LIBs but also advances the field towards achieving high-performance, sustainable, and cost-effective energy storage solutions. The insights gained and the methodologies developed lay a solid foundation for future research and development in battery technology, steering the global shift towards electrified transport and renewable energy sources.

506a Salt Hydrate Eutectics: Tailored Design of Equilibria, Reversibility, and Stability in Complex Phase Change Material Systems

Patrick Shamberger

Salt hydrate phase change materials (PCMs) represent a promising class of materials for thermal energy storage in buildings and environmental climate control systems, but are currently limited by the lack of compounds with reversible melting-solidifying behavior within particular temperature ranges of practical use for building thermal energy storage applications (approximately 10 to 30 °C). Here, I will present a general strategy, harnessing the ability of salt hydrates to form pseudo-binary or higher-order eutectics, to refine the melting point of low-cost stable eutectic salt hydrate PCMs for building thermal energy storage applications. While this approach utilizes a well-documented strategy of relying on a thermodynamically invariant point (i.e., the eutectic point) to result in well-defined and robust melting behavior, it introduces a number of unique challenges, including 1) efficient exploration of multi-component systems and identification of eutectic compositions, 2) nucleation in multicomponent salt hydrate systems, including the potential impact of metastable liquids on the stability of such eutectics, and 3) complex interactions in PCM materials systems, which generally include multiple distinct phases or chemical constituents added to the system to address nucleation, heat transfer, and to encapsulate the salt hydrates. In such systems, interactions between different components can serve to aid or to deteriorate the overall performance of the material system. Throughout, examples will be provided based on nitrate and chloride hydrate systems and their eutectics.

Efficient exploration of salt hydrate compositions is required because of the vast compositional space that these systems can span. Here, we utilized the modified Brunauer–Emmett–Teller (BET) method as an efficient predictive method, which requires minimal initial knowledge on material systems. We use this method to evaluate pseudo-binary and ternary eutectics formed between five nitrate hydrates (of Li, Mg, Ca, Mn, and Zn), and three anhydrous nitrates (of Na, K, and NH_4). We demonstrate the overall success of the method, while highlighting some of its limitations, including the inability to predict isomorphous solid solutions, and challenges faced in describing anhydrous nitrate phases.

A number of salt hydrates are known to experience nucleation-limited undercooling, although the degree of this phenomenon varies widely, even across chemically similar salt hydrates. Here we address fundamental questions related to the occurrence of undercooling in multi-component salt hydrates. In particular, we demonstrate the intrinsic variability of undercooling observed in different related salt hydrate phases, which illustrates the difficulty in predicting nucleation behavior in multi-component systems. Furthermore, while inclusion of a single class of nucleation particles (targeting a single crystalline phase) can serve to promote complete solidification of multiple crystalline solids in a pseudo-binary eutectic, they do not necessarily initiate crystallization simultaneously. This can result in a potential for phase segregation, even in a system that (in equilibrium) solidifies in a manner that would not be susceptible to phase segregation. Finally, we illustrate the challenge in identifying active nucleation particles in reactive salt hydrate systems, drawing from a pathological example in the $CaCl_2 \cdot 6H_2O$ system, where Ba-based compounds undergo multiple cation-substitution and precipitation reactions, obscuring the nature of the active nucleation particle.

Finally, I will address the topic of PCMs as systems of materials, and will introduce some of the interactions that can occur between different constituents of that system, including 1) the PCM, 2) nucleation particles, 3) encapsulating materials, 4) high thermal conductivity graphitic additives, and 5) polymer molecules introduced to thicken or form a gel. Specifically, some of these components can serve multiple functions (e.g., encapsulating materials which also nucleate solidification), while other of these components can interact in a negative destructive manner (e.g., polymer gel molecules, which can reduce the efficacy of nucleation particles in some cases). Additional examples include the potential effect of shape stabilization on the cyclic degradation observed in thermal conductivity due to cycles of expansion and contraction, as well as the effect of polymer networks to limit complete compositional mixing during melting cycles, which can result in formation of bimodal or trimodal characteristic melting peaks.

Overall, this presentation seeks to emphasize the salient features of salt hydrate eutectic systems, while providing illustrative examples from nitrate and chloride eutectics. We will conclude by identifying remaining challenges in the field, as well as instructive strategies to continue to advance low cost salt hydrate thermal energy storage materials.

506b Enhancing Building Thermal Management with Thermochemical Materials-Based Thermal Energy Storage

Alondra Perez, Andrew Martin, Chris Dames, Sumanjeet Kaur

Buildings globally are significant contributors to energy consumption, with the United States playing a prominent role in this regard. Among the various energy demands, thermal loads stand out for their substantial contribution to CO2 emissions. Addressing these loads is pivotal for enabling deeper integration of renewable energy sources. On-site Thermal Energy Storage (TES) emerges as a crucial solution for managing peak loads and mitigating the intermittency of renewable energy sources. Thermochemical Materials (TCMs), comprising reactive pairs of inorganic salts and water vapor, offer promising advantages. Notably, they boast a higher theoretical energy density of approximately 500 kWh/m3 and demonstrate minimal self-discharge due to energy storage within chemical bonds. These characteristics make them ideally suited for compact, stand-alone solutions catering to daily-seasonal energy storage within buildings. However, TCMs encounter challenges, particularly regarding material instabilities at both the particle and reactor levels (packed beds of salt), resulting in suboptimal multi-cycle efficiency and elevated levelized costs of storage.

In conventional salt hydrates, water can comprise approximately 1/4–1/2 of the mass of the salt hydrate itself. For instance, in commonly investigated salts like MgSO4·7H2O, water occupies approximately 51% of the salt hydrate's mass, leading to significant volume and porosity changes during dehydration (approximately 71.8% volume reduction).

Previous investigations into the transitions of TCM salt hydrates between various hydrate phases have revealed different mechanisms during dehydration and hydration. While hydration may follow different pathways, such as direct solid–solid transition or dissolution and recrystallization processes, salt dehydration for most salts typically involves the diffusion of water molecules out of the salt hydrate crystal. Consequently, this water removal induces solid-state transformation, causing mechanical stress and strain on the crystal, leading to defects such as dislocations and cracks, ultimately resulting in salt pulverization. While both hydration and dehydration may contribute to the deterioration of salt hydrates over cycling, dehydration is considered the primary contributor to salt hydrate pulverization.

We have developed a model to predict the pulverization limit (i.e., critical size, Rcrit) of salt hydrates and validated it for various salt hydrates. Additionally, we demonstrated the effect of Rcrit on dehydration and hydration kinetics, providing insights into the long-term stability of salt hydrates and their composites. To determine the extent of salt hydrate pulverization due to expansion during cycling, we developed a theoretical model for a single spherical particle by solving coupled mechanical stress and mass diffusion equations. While this model assumes a perfectly spherical, defect-free particle with isotropic material properties, the final predictions are minimally affected by these assumptions, focusing instead on the rate of dehydration.

Based on our model, we optimized the synthesis recipe for TCM composites, preconditioning pristine salts to Rcrit before mixing them with the host matrix. TCM composites made using preconditioned salt hydrates exhibited mechanical integrity and stability for over 2000 cycles, compared to 20 cycles with as-received salt hydrates. Preconditioning salts minimize crack formation, slipping, and changes in expansion/contraction behavior within the host matrix, resulting in more predictable mechanical behavior during cycling and potentially better long-term performance of composite TCMs. Furthermore, we are developing correlations between material properties and reactor performance to identify pathways for enhancing heat and mass transport efficiency at the reactor level.

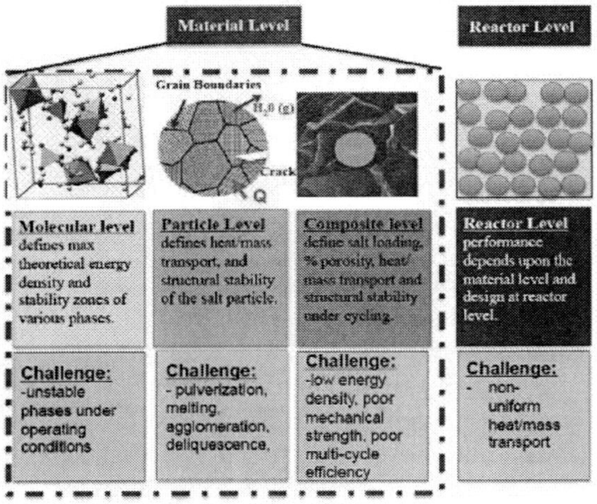

TCMs suffer from instabilities at the material (salt particles) and reactor level (packed beds of salt), resulting in poor multi-cycle efficiency.

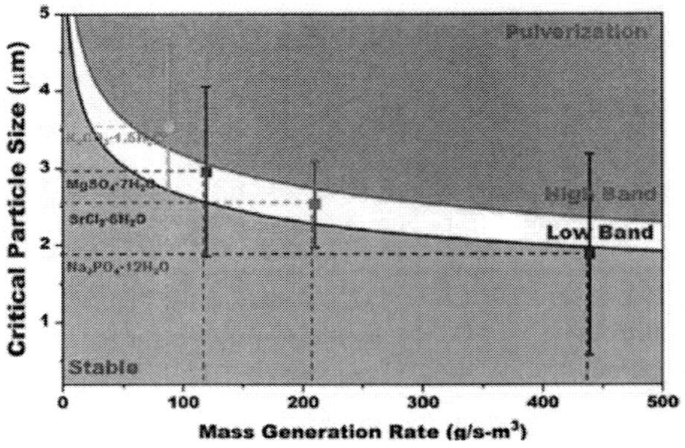

Model along with experimental data on the critical particle size of salt hydrates. Two set of material constant values were taken to represent the range of mechanical and intrinsic properties for various salt hydrates.

Martin, A., Lilley, D., Prasher, R. & Kaur, S. Particle Size Optimization of Thermochemical Salt Hydrates for High Energy Density Thermal Storage. *Energy Environ Mater* (2023) doi:10.1002/eem2.12544.

506c Performance Analysis of Fluidized Bed Reactors for Thermochemical Energy Storage Based on the $CaO/CaCO_3$ and $MnAl_2O_4/MnAl_2O_{4-\Delta}$ Systems

Maria Anna Murmura, Valentina Biagioni, Antonio Brasiello, Maria Cristina Annesini

Thermochemical energy storage is an interesting tool to decouple the intermittent production of electricity from renewable sources from its use. The process is based on gas-solid reactions characterized by high heats of reaction: the endothermic reaction is carried out when excess energy is available (charge phase), while the exothermic reaction is allowed to take place when the energy demand is higher than the production (discharge phase).

The efficiency of these storage systems and the possibility of coupling them with downhill energy-demanding processes depend on both the characteristics of the material employed, such as heat of reaction, kinetics, and cyclability, and the reactor design. More specifically, the system should be capable of guaranteeing a constant outlet temperature of the gas during the discharge phase for long periods of time. Both the $CaO/CaCO_3$ and the $MnAl_2O_4/MnAl_2O_{4-\delta}$ materials have been found to be promising for application in thermochemical storage reactions. To improve the thermal management of the process, the use of fluidized bed reactors has been proposed, because of the high heat transfer rate between solid and gas and their capability of maintaining almost isothermal conditions; however, these reactors are less flexible in terms of operating conditions because of the need to maintain adequate fluidization conditions. These are, in turn, strongly affected by changes in density of the solid particles that can take place during gas-solid reactions. Consequently, it is important to tailor the operating conditions and reactor geometries to the specific system in use and to ensure that the fluidization regime is maintained throughout the reaction.

In this work, a model for a fluidized bed reactor operating with either the $CaO/CaCO_3$ or the $MnAl_2O_4/MnAl_2O_{4-\delta}$ was developed. The effect of the degree of advancement of the reaction on the fluidization characteristics was analyzed and simplified models were developed to evaluate key performance indicators at a low computational cost.

506d Dynamic Simulation and Optimization of an Adsorption Heat Storage

Alix Untrau, Lorenz Biegler, Frédéric Marias, Jean-Michel Reneaume, Sylvain Serra, Sabine Sochard

The energy transition requires the extensive use of renewable energies, including those from intermittent sources such as solar and wind energies. For heat consumption in buildings and in industrial processes, solar thermal heat represents a good alternative to fossil fuels. However, the production and consumption of heat are variable and asynchronous. A thermal storage is therefore required to satisfy the heat demand. Adsorption heat storage is a promising technology since it presents a high energy density ensuring its compacity. Moreover, the heat is stored as a chemical potential, resulting in negligible heat losses. In an adsorption storage connected to a solar thermal system, the solar excess heat is used to desorb the molecules (endothermic regeneration of the adsorbent) and the exothermic adsorption is performed when heat needs to be recovered. Currently, adsorption heat storage is not competitive against sensible or latent heat storage due to the high cost of material. Nonetheless, the choice of adsorbent/adsorbate couple, the design of the adsorption column as well as the operating parameters impact the energy density of the storage (Lefebvre and Tezel, 2017). Thus, optimizing the design and operation of the adsorption heat storage could help improve its economic viability.

This study aims at modeling and optimizing an adsorption heat storage to show its ability to satisfy the heat demand of a building with renewable heat. For integration in a building, an open loop system with water chosen as the adsorbate ensures a simpler structure. The adsorbent chosen is a zeolite 13X. The design of the system is retrieved from an experimental pilot in (Tatsidjodoung *et al.*, 2016).

The model used was built on a previous detailed model developed by (Ferreira *et al.*, 2021), and validated with the experimental data from (Tatsidjodoung *et al.*, 2016). The model is based on mass balances on the fluid components, air and water, and on energy balances on the fluid and the porous media. A Linear Driving Force (LDF) model (Glueckauf, E., 1955) is used to represent the adsorption process, including the different mass transfer resistances: external, macroporous and microporous. The equilibrium model chosen is the Aranovich–Donohue model as formulated by (Ahn. H., 2019). The differential-algebraic system of equations is solved with ode15s in MATLAB. Some simplifying assumptions have been implemented to obtain a simple model suitable for optimization: constant thermophysical properties (except for the fluid density), negligible axial dispersion and fluid and solid at the same temperature. Moreover, the spatial discretization scheme has been improved and finite volumes are considered to ensure mass conservation.

The results of the dynamic simulation of an adsorption phase show the impact of the operating parameters on the energy released. The duration of the phase, the temperature lift obtained, and the overall quantity of energy recovered depend on the inlet fluid temperature, humidity, and velocity, as well as the size of the storage. It seems that a dynamic optimization could help to choose the best operating parameters to satisfy a dynamic heat demand at the lowest cost. In future work, the size of the storage tank as well as the operating parameters will be optimized in some case studies for the adsorption phase. The integration of the adsorption heat storage into a solar thermal house heating system will then be implemented and the overall system optimized.

References

Ahn, H., 2019, Equilibrium theory analysis of thermal regeneration of a humid adsorption column: Selection of optimal hot purge gas temperature, *Chem. Eng. Res. Des.*, 151, 91–99.

Ferreira, S., Sochard, S., Serra, S., Marias, F., Reneaume, J.-M., 2021, Modelling of an Adsorption Heat Storage System and Study of Operating and Design Conditions, *Processes*, 9, 1885.

Glueckauf, E., 1955, Theory of chromatography. Part 10—Formulæ for diffusion into spheres and their application to chromatography. *Trans. Faraday Soc.*, 51, 1540–1551

Lefebvre, D., Tezel, F. H., 2017, A review of energy storage technologies with a focus on adsorption thermal energy storage processes for heating applications, *Renewable and Sustainable Energy Reviews*, 67, 116-125

Tatsidjodoung, P., Le Pierrès, N., Heintz, J., Lagre, D., Luo, L., Durier, F., 2016, Experimental and numerical investigations of a zeolite 13X/water reactor for solar heat storage in buildings, *Energy Convers. Manag*, 108, 488–500.

506e Salt Hydrate Composites with Improved Cyclability for Thermochemical Energy Storage in Buildings

Madeline Morrell, Erik Barbosa, Akanksha Menon, Ahmed Mahfouz

Fossil fuels are widely used for heating, especially in the buildings and industrial sectors. In the United States, 60% of the energy consumption in the residential buildings sector is attributed to thermal loads such as space heating and hot water. Heat decarbonization is thus critical, motivating the development of a thermal battery that can be charged with low-cost electricity or solar energy and can discharge zero-carbon heat to meet thermal loads on demand. Thermochemical reactions can be leveraged for heat storage and release given their high volumetric energy densities and negligible self-discharge for long duration applications. Thermochemical materials (TCMs), such as salt hydrates, undergo reversible solid-gas chemical reactions at temperatures below 100°C that make them suitable for integration with buildings. The charging reaction is an endothermic reaction that dehydrates the salt, whereas the discharging reaction is an exothermic reaction where the salt hydrates. For a practical application, TCMs must be able to withstand hundreds of charge-discharge cycles, however literature has demonstrated that significant structural changes limit the reversibility of the reaction. For example, $CaCl_2$ experiences a 60% volume reduction as it dehydrates from its hexahydrate to anhydrous state. This cycling-induced stress concentration leads to material pulverization, decreasing energy storage capacity after only a few cycles. In addition to structural changes, the hydration reaction can cause deliquescence, which then leads to agglomeration when the salts are dehydrated. This hygrothermal instability increases water vapor diffusion resistance, which reduces the kinetics and energy density of these materials.

To address these challenges, we develop composite architectures that reduce thermo-mechanical degradation of salt hydrates and subsequently enhance thermal battery performance. Two approaches are demonstrated to improve material cyclability: *(i)* salt encapsulation within a porous polymer matrix, and *(ii)* salt impregnation within mesoporous silica. In the first approach, cyclability is improved by synthesizing different composites that improve the mechanical properties of salt hydrates, such as $CaCl_2$, by incorporating it within a hydrogel matrix with elasticity, such as alginate and polyacrylamide. To address agglomeration, in the second approach $CaCl_2$ is nanoconfined within porous silica that shifts the hydration behavior by leveraging capillarity. The energy density of the different composites is characterized and the hygrothermal stability over multiple hydration and dehydration cycles is investigated. The hydration kinetics of these salt composites are evaluated through measuring the water uptake using a TGA-DSC for various relative humidity and temperature settings. It is then possible to compare various salt composite hydration extents through obtaining experimental kinetic constants of these salts and computationally modeling the reaction advancement using solid-gas reaction models. Overall, this study compares a variety of composite salt hydrates, and assesses the coupled relationship between their mechanical, kinetic, and thermophysical properties to achieve materials that can significantly enhance long-term energy storage performance for building applications.

506f Degradation and Recovery of the Efficacy of Strontium Chloride Hexahydrate As a Nucleation Particle for Calcium Chloride Hexahydrate

Denali Ibbotson, Patrick Shamberger

As alternative renewable energy sources, which tend to be intermittent and in some cases unpredictable in nature, become more widespread, the need for technologies to store surplus energy for use at a later time will only grow. Thus, to move towards full decarbonization of the energy grid, effective thermal energy storage systems could play an important role in this process. Specifically, phase change materials (PCMs) can passively store and release building process heat, either assisting with cooling or heating of internal building air, thereby introducing a means of displaced peak energy load for buildings. However, robust and economical phase change materials that are tailorable to a specific desired energy storage temperature remain limited, and still face challenges with robust long-term thermal behavior. Salt hydrates represent one class of PCMs of interest due to their high volumetric energy densities, low cost, and the ability to tailor the thermal energy storage temperature through the design of custom salt hydrate eutectics. However, they can also experience undercooling, a nucleation-limited phenomena which results in the existence of a metastable liquid below the melting point. This phenomena can lead to undesired issues, including incongruent melting and phase segregation. To combat these issues, nucleation particles are often added to systems to induce nucleation and suppress formation of metastable phases, thus mitigating the chance of phase segregation occurring.

While compounds that are isostructural to the desired PCM tend to have low interfacial energy with the target phase, resulting in low degrees of undercooling, these phases (which also tend to be salt hydrates) may not be stable under all conditions, reacting to form lower hydrates, or even anhydrous solids. In this study, we investigated the efficacy of strontium chloride hexahydrate ($SrCl_2 \cdot 6H_2O$; SCH) as a nucleation particle to reduce undercooling in calcium chloride hexahydrate ($CaCl_2 \cdot 6H_2O$; CCH) over a range of temperatures and concentrations. Calcium chloride hexahydrate is a stoichiometric salt hydrate of interest for use in thermal energy storage systems because it is a low-cost salt hydrate that melts at 29°C [1, 2]. However, calcium chloride hexahydrate is also prone to undercooling and phase segregation due to the presence of a peritectic point at 29.45°C, where the metastable state, calcium chloride tetrahydrate ($CaCl_2 \cdot 4H_2O$) can also be present. Therefore, nucleation particles are often added to CCH in order to induce nucleation, suppress the formation of a metastable state, and to reduce undercooling. Throughout many different studies, strontium chloride hexahydrate is often noted as an effective nucleation particle for CCH due to their isostructural relationship [3, 4].

Here, we demonstrate that the effectiveness of SCH as a nucleation particle is dependent on the time-temperature history of the system, suggesting a reaction pathway that results in the degradation of the desired SCH phase. For example, increasing the temperature of a system of CCH mixed with SCH to greater than ~65°C for a minimum of two minutes increases the degree of undercooling of the system, indicating that SCH can lose efficacy. Previous solubility studies explain this behavior, as it has been previously reported that CCH and SCH cannot exist in equilibrium with each other at specific temperatures and concentrations, and other hydrate phases of SCH will instead be present [5]. Surprisingly, cooling below ~ -30°C recovers this efficacy, but only if the system is cooled for at least 5 minutes at this cold temperature prior to holding at 0°C. Thus, if the system is allowed to solidify completely, we are able to regain the phase that is responsible for nucleation of CCH, which indicates that the reactivity of SCH can affect its efficacy as a nucleation particle.

In contrast, reactivity between a salt hydrate PCM and a nucleation particle can be harnessed in a positive way, resulting in in-situ reactions with products that are both stable, and active nucleators. We will introduce an example of this mechanism describing recent results from Ba-based compounds which undergo in-situ reactions upon exposure to liquid CCH, resulting in a decrease in undercooling of ~10°C. However, the Ba-based nucleation particles have no apparent structural relationship to CCH, instead we hypothesize that they are effective nucleators due to the existence of atomic-scale defects on the surface of a solid solution phase. Therefore, in the case of strontium chloride hexahydrate, the strategy of isostructural nucleation particles does present a path towards effective nucleation, however the reactivity of the system must also be considered and evaluated.

References:

[1] G. A. Lane, *Solar Heat Storage: Latent Heat Materials Volume II: Technology.* CRC Press, 1986.

[2] B. Carlsson, H. Stymne, and G. Wettermark, 'An Incongruent Heat-of-Fusion System - CaCl2.6H2O - Made Congruent through Modification of the Chemical Composition of the System,' *Solar Energy,* vol. 23, pp. 343-350, 1979.

[3] L. Yan, J. Charles, C. E. Romero, S. Neti, and G. Balasubramanian, 'Optimizing supercooling and phase stability by additives in calcium chloride hexahydrate for cyclical latent heat storage,' *International Communications in Heat and Mass Transfer,* vol. 149, 2023, doi: 10.1016/j.icheatmasstransfer.2023.107119.

[4] N. Kumar and D. Banerjee, 'Thermal Cycling of Calcium Chloride Hexahydrate With Strontium Chloride as a Phase Change Material for Latent Heat Thermal Energy Storage Applications in a Nondifferential Scanning Calorimeter Set-Up,' *Journal of Thermal Science and Engineering Applications,* vol. 11, no. 5, 2019, doi: 10.1115/1.4042859.

[5] G. O. Assarsson and A. Balder, 'Equilibria between 18 and 114 in the aqueous ternary system containing Ca2+, Sr2+, and Cl-,' 1953.

506g Development of Shape-Stabilized Salt Hydrates with Polymer Components for Low-Cost Thermal Energy Storage

Sophia Ahmed, Kartik K. Rajagopalan, Svetlana A. Sukhishvili, Patrick Shamberger

Salt hydrates are a class of phase change materials (PCMs) that are of interest due to their lower cost-energy density and higher volumetric energy density than other PCMs such as paraffins. This PCM also has the ability to form eutectics, like other PCMs, which allows for the chemical specification of bespoke PCMs designed to store thermal energy at specific invariant temperatures. Unfortunately, the use of salt hydrates in thermal energy storage (TES) is hindered by fluid flow and overall redistribution when the PCM is in its liquid state, as well as their susceptibility to form a metastable liquid state during cooling. Additionally, salt hydrate eutectics are potentially susceptible to phase segregation if solidification occurs in a metastable manner, as density-driven differences preferentially causes one phase or more to separate out. This segregation would result in inhomogeneity throughout the material. Here, we investigate the use of polymer gel networks, referred to as salogels due to the high salt concentrations, where a small amount (<5 wt%) of a compatible polymer is incorporated in a salt hydrate system, resulting in a matrix formed through hydrogen-bonding and polymer-ion interactions. The resultant polymer network enables the PCM to retain its shape above its melting temperature, T_m, by preventing its flow, allows the solid-to-liquid phase transition of the salt hydrate that provides the desired high latent heat, and can potentially mitigate phase segregation by trapping all eutectic components in place even as the PCM becomes metastable below T_m. In addition, these salogels are thermo-reversible as they are physical gels, meaning the salogel can flow above its gel-to-sol transition temperature (T_{gel}) allowing processability, and it exists as a stiff solid below T_{gel}. However, limited data is available on the thermal cycling behavior of salogels made with stoichiometric salt hydrates and their eutectics, including the effect of the salogel polymer framework on the efficacy of nucleation particle (NP), which serve to nucleate the crystalline salt hydrate from their exposed solid surfaces. Our previous work investigated the cycling and nucleation behaviors of pure salt hydrates and their family of eutectics with and without any NPs in the absence of polymers. Now, our work seeks to determine how a polymer component can affect the cycling and nucleation behavior in a system, especially when it is expected that the polymer plays no active role in the melting behavior of the PCM.

Here we present a family of zinc nitrate hexahydrate ($Zn(NO_3)_2 \cdot 6(H_2O)$) salogels, including those made with pseudo-binary eutectics ($Zn(NO_3)_2 \cdot 6(H_2O)$-$NaNO_3$, $Zn(NO_3)_2 \cdot 6(H_2O)$-KNO_3, and $Zn(NO_3)_2 \cdot 6(H_2O)$-NH_4NO_3). Through the addition of a low concentration of polyacrylamide (PAAm), we observe the effects of this polymer addition on the cycling and nucleation behavior of these systems, as both have plagued salt hydrate eutectics. In prior work, the base, polymer-absent $Zn(NO_3)_2 \cdot 6(H_2O)$ systems did not undergo phase segregation as a result of metastability upon cycling; the presence of a polymer should not change this at all. From our work, we have seen that salogel systems with little to no anhydrous salt concentration ($Zn(NO_3)_2 \cdot 6(H_2O)$ and $Zn(NO_3)_2 \cdot 6(H_2O)$-$NaNO_3$) show consistent, uni-modal melting behavior across >150 thermal cycles even as they were cycled to maximum temperatures near T_{gel}. As anhydrous salt concentration increases in a eutectic system ($Zn(NO_3)_2 \cdot 6(H_2O)$-KNO_3 and $Zn(NO_3)_2 \cdot 6(H_2O)$-NH_4NO_3), the melting behavior devolves into a multi-modal (bi- or tri-modal) shaped melting curve across >150 cycles if the maximum cycling temperature is in proximity of T_{gel}. However, this shape appears to stabilize after a number of cycles. This is exceptionally seen in the $Zn(NO_3)_2 \cdot 6(H_2O)$-NH_4NO_3

salogel as bimodal melting curves are observed as the system is cycled close to its T_{gel} of ~30 °C, show that the loading of the salogel into any PCM module above T_{gel} can have potentially negative long-term impact on its melting behavior. The bimodal behavior is potentially due to the PAAm component swelling more at higher temperatures, resulting in compositional changes to the PCM. This is seen in polarized light microscopy studies as it was observed that precipitates formed at higher temperatures.

For the base $Zn(NO_3)_2 \cdot 6(H_2O)$ system, talc has been identified as an effective NP. In neat salogel systems, undercooling decreased to <15 °C even in the absence of an NP when compared to the polymer-absent form, which had undercooling values <25 °C, showing that the polymer aids in nucleation. However, talc usage is warranted as it makes undercooling low and consistent across cycles. The work presented here showcases the effects a polymer component can have on the cycling and nucleation ability of salt hydrate systems, especially as its need arises when trying to make shape-stable TES media. In terms of undercooling, the polymer reduces it in salt hydrate systems, maximizing more volume available for PCMs. For melting, the polymer component in some systems does not hinder its melting behavior when compared to its polymer-absent form; some systems did experience changes at higher temperatures after polymer was added, posing a potential hinderance in implementation as a TES media.

665a Analysis of Cathode-Electrolyte Interfaces in Lithium-Ion Batteries

Meng Zhou, Hongmei Luo

Lithium-ion batteries (LIBs) have been widely used in portable electronics and electric vehicles. The future development and improvement of LIBs require a better understanding of electrochemical activities at battery electrode/electrolyte interfaces. The formation of interfacial layers on the anodes has been extensively studied, however, the formation of cathode-electrolyte interfaces (CEI) layers may also accelerate LIB degradation. Therefore, the characterization of CEI is critical for understanding the interfacial processes, including transition metal dissolution, oxygen release, and electrolyte oxidation. In this presentation, we will focus on characterizing oxidative processes occurring at Pt, glassy carbon, and model LIB cathode materials by scanning electrochemical microscopy. Electrochemical measurement analysis indicates multiple anodic reaction processes occurring at Pt and glassy carbon at high voltage due to the electrolyte oxidation. The electrolyte oxidation processes generate reducible products from metal oxide cathodes that are detected at different voltage ranges, which can be associated with the metal dissolution of the electrochemically active materials from cathodes. We propose a preliminary model mechanism that connects transition metal dissolution with catalytic electrolyte oxidation and potential interfacial film formation. Our observation offers insights into the CEI formation and evolution mechanism, providing information on LIB interfacial processes associated with cathode degradation.

665b Design of Polymer Interfaces for Lithium Metal Batteries Via Physical and Chemical Interactions

Rachel Huang, Yi Cui, Zhenan Bao

LMBs are widely regarded as the next generation of energy storage devices because of their high anode specific energy density of 3860 Ah kg^{-1}. However, the long-term operation of LMBs is hindered by the uncontrolled deposition process of Li due to the unstable solid electrolyte interphase (SEI) formed when Li contacts the electrolyte. The key to improving battery cycling performance is stabilizing the Li electrode.

Polymer coatings on Li can address issues related to each step of heterogenous Li deposition. In this talk, I will highlight three mechanisms by which polymer coatings can improve lithium metal anode performance: 1) maintaining uniform coverage to prevent pinholes; 2) homogenizing charge distribution; and 3) regulating reactions at the interface to improve the SEI chemistry. I will dive deep into how each mechanism is realized by molecular level design. Lastly, I will discuss the performance of polymer coatings when paired with different electrolytes and lay out future directions towards designing polymers for advanced electrolytes.

1. Z Huang et al, *Effects of Polymer Coating Mechanics at Solid-Electrolyte Interphase for Stabilizing Lithium Metal Anodes*, *Adv. Eng. Mat*, 2022
2. Z Huang et al, *A Cation-Tethered Flowable Polymeric Interface for Enabling Stable Deposition of Metallic Lithium*, *JACS*, 2020
3. S Choudhury*, Z Huang*, C Amachuku*, P Rudnicki*, et al, *Ion Conducting Polymer Interfaces for Lithium Metal Anodes: Impact on the Electrodeposition Kinetics*, *Adv. Eng. Mat*, 2023
4. Z Huang et al, *A salt-philic solvent-phobic interfacial coating design for lithium metal electrodes*, *Nat. Energy* , 2023

665c Self-Discharge of Layered Oxide Cathodes Induced By Carbonate-Mediated Hydrogenation

Gang Wan

Significant efforts have been made to improve the energy and power characteristics of cathodes in Li-ion batteries, while the research into cathode self-discharge has lagged behind. Self-discharge is the voltage drop experienced by all rechargeable batteries while stored in the charged state. The voltage drop becomes more problematic with increased cathode voltage and temperature. The physicochemical nature of internal reactions leading to cathode self-discharge is hardly understood and seldom discussed.

To tackle this challenge, the structure and redox evolution of commercial $LiNi_{0.5}Mn_{0.3}Co_{0.2}O_2$ electrodes and single crystal cathode thin films upon the self-discharge in carbonate-based electrolyte is revealed by surface-sensitive X-ray scattering, spectrometric and electrochemical characterizations in conjunction with thermodynamic analysis. As evidenced by the interfacial toolkits and theoretical calculations, there is an evolution and growth of cathode surface reduction layer in LP57 electrolyte after self-discharged from different potentials. Structural and chemical chacterizations as well as the elemental depth profiles within cathode thin film confirms proton-insertion-induced layered cathode hydrogenation. Calculation shows that such process is both thermodynamically and kinetically favorable and is triggered by the interfacial hydrogen atom abstraction of methylene group in carbonate solvent on delithiated cathode surface. A combination of experimental and theoretical studies reveals the carbonate-mediated cathode hydrogenation mechanism accounting for the voltage drop in the self-discharge. This offers additional understanding regarding defect generation and the degradation mechanism in layered cathodes beyond the traditional behaviors of lithium-diffusion-induced self-discharge and rock-salt phase induced cathode degradation. This study offers foundational knowledge of the interfacial degradation of cathode that can be translated to the rational design of improved cathodes and electrolytes, and the self-discharge mechanism studies in other electrochemical ion insertion materials and devices.

665d Insights into over-Lithiation Mechanisms and Electrochemical Restoration during Lithium-Ion Battery Direct Recycling

Maura Appleberry, Zheng Chen

Direct recycling is a promising alternative for low-cost and more environmentally friendly disposal of lithium-ion batteries. Typically, direct recycling methods involve a 'relithiation' step to address the accumulated Li loss during electrode use, followed a short annealing step to resolve remaining surface degradation. It is widely reported that excess Li is necessary during the annealing step to prevent further degradation. Our work has built on previous advancements of optimizing hydrothermal relithiation parameters, and we have demonstrated that it is not necessary to pre-determine added Li at each stage, making this approach lucrative for scaling. However, we have observed the possibility of 'over-lithiation' for various lithium nickel cobalt manganese oxide (NCM) cathode materials, where the lithium stoichiometry exceeds 1.05. This over-lithiation can lead to the broadening of cell parameters in layered transition metal oxides under certain conditions. Through a study of over-lithiation (Li $\leq$1.2), the crystalline phases of excess Li within the particles are identified and their resulting impact on electrochemical performance is measured. Furthermore, approaches to remove excess Li to return stoichiometric balance and optimal electrochemical performance are proposed. This improved understanding of over-lithiation mechanisms and lithium diffusion enables the full restoration of pristine electrochemical capacity during direct recycling, thus paving the way for at-scale rejuvenation.

665e Ultrafast Electrodeposition of Faceted Li Metal

Yuzhang Li

Electrochemical deposition of metals with precise morphological and chemical control is of fundamental importance to a broad range of electrochemical systems, including energy storage, large scale manufacturing, electronics, and electrocatalysis. For non-reactive metals (e.g. Ni, Pt), electroplating is a commercially established process that gives the desired control of deposition morphology and chemistry. However, deposition of reactive metals (e.g. Li, Na) critical for enabling next-generation batteries is complicated by metal surface corrosion that occurs concurrently with metal deposition, making such electrodeposition processes neither well-understood nor well-controlled. This work highlights the electrodeposition of lithium (Li), a reactive metal anode plagued with uncontrolled dendritic morphologies and surface corrosion, both of which are intimately linked and occur simultaneously. Despite tremendous research efforts to mitigate these issues, common literature descriptions of Li metal deposits (e.g. "chunky", "pancake") and their corrosion films (e.g. "strong", "stable") are ill-defined and often qualitative, highlighting our incomplete understanding of the deposition process and hindering the practical application of high energy battery chemistries enabled by Li metal. Our work seeks to bridge this critical gap in understanding by solving the fundamental challenge of (1) decoupling simultaneous electrodeposition from surface corrosion of reactive metals and (2) observing their reactive nanoscale interface.

665f Sacrificial Presodiation Additive Toward High-Capacity Sodium All-Solid-State Batteries

Dapeng Xu

Sodium all-solid-state batteries (NaSSBs) are garnering interest due to their sufficient element resources and superior safety features. Using an alloy-type anode in the full cell configuration, such as tin (Sn) and antimony (Sb), can substantially enhance capacity and energy density comparing to state-of-art hard carbon anode. Additionally, due to the high Na+ diffusivity, the alloy and de-alloy process do not require mixing with electrolytes to enhance ionic conductivity, which reduces inactive component in NaSSBs. However, the irreversible loss of Na^+ at the alloy anode side during the initial cycle results in diminished capacity and stability, impairing the battery's overall performance. This study presents an easy-to-implement cathode presodiation strategy that effectively mitigates Na^+ loss with an additive in the cathode and improves overall energy density. Na2S was used as the main component in pre-Na agent due to the large theoretical capacity (~700mAh/g) and matched Na+ ion extraction voltage (~1.8V vs. Na/Na+). Other components of Na_3PS_4 (NPS) and acetylene black (AB) were used to activate Na2S by introducing ionic and electronic conductivity. The optimized composition of pre-Na agent (Na2S:NPS:C=2:1:1) delivers a high capacity of 550mAh/g with a 90% utilization of its theoretical capacity,. Material characterization reveals the mechanism of the sodium supplement process: chemical composition and electrochemical changes during charging. The cell performance with $NaCrO_2$ (NCO), $Na_{0.55}Ni_{0.1}Fe_{0.1}Mn_{0.8}O_2$ (NNFM) cathode, and Sn, Sb anode shows good material compatibility, promoted cycling performance and interfacial stability. This work goes beyond previous efforts by first providing a scalable, accessible, and efficient method to enhance the full-cell performance of NaSSBs significantly, advancing the technology of solid-state batteries, and offering a promising avenue for sustainable and efficient energy storage.

665h Electrolyte Engineering Enables High-Performance Span-Based Lithium-Sulfur Batteries Under Practical Conditions

Jianjun Chen, Yuqing Fu, Juchen Guo

Lithium-sulfur (Li-S) batteries are emerging as a highly promising candidate for next-generation energy storage systems, distinguished by their impressive energy output, cost-effectiveness in materials, and environmental benignancy characteristics. Lithium polysulfides, as the discharge intermediates, show great reaction activity towards Li anode which result in continuous sulfur loss and Li corrosion, contributing to capacity decreased and poor cycle stability. Applying chemical sequestration that covalently bond sulfur to a carbonaceous host presents as a promising alternative to eliminate polysulfides generation. Among them, sulfur-polyacrylonitrile (SPAN) stands out as one of the most promising materials. Li-SPAN batteries have been demonstrated with stable cycling up to hundreds of cycles which is benefiting from the negligible polysulfides dissolution that react with carbonate-based electrolytes to form a stable cathode electrolyte interface (CEI). However, challenges remain in maintaining high performance under high loading and lean electrolyte conditions because of the poor compatibility between carbonate-based electrolytes and Li anode.

In this work, we first distinguish ethylene carbonate (EC) as the optimized selection among carbonate solvents for constructing the robust CEI. In order to get insight into the behind mechanism, kinetic study and galvanostatic intermittent titration technique (GITT) combined with electrochemical impedance spectroscopy (EIS) technology using three-electrodes pouch cells are developed. Chemical analysis like solid NMR and surface analysis such as XPS are applied to study the components of CEI formed in different carbonate electrolytes. Furthermore, as addressing the compatibility issue between carbonates and Li anodes, ether and fluorinated ether solvents are introduced with improved columbic efficiency. Befitting from the robust CEI and SEI, a Li-SPAN pouch cell with high loading of 7 mg cm^{-2}, thin Li anode of 100 μm and lean electrolyte condition of 2.5 μl mg^{-1} was demonstrated with stable stability of 150 cycles and deliver a high energy density of 170 Wh kg^{-1} (based on whole system).

166a Importance of Transport in Fuel Cells and Their Components

Adam Z. Weber

As electrochemical technologies become increasingly important in our energy paradigm, especially fuel cells, there is a need to examine them holistically. Furthermore, for such technologies to become practical, they need to operate at high current densities to minimize various cell costs. This operating space necessitates the need for efficient transport of reactants and removal of products from the reaction site and in a solid-state architecture that utilizes ion-conducting polymers (ionomers). In this talk, we will explore the various tradeoffs endemic in such gas-diffusion electrode (GDE) architectures undergoing fuel-cell reactions. We will examine the major phenomena within fuel cells with a focus on water management, as well as examine changes that occur during operation (e.g., increased wettability, catalyst particle growth) that exaggerate the local transport limitations. Such tradeoffs are quantified through multiphysics modeling and key diagnostics of the cells and components including breakdowns of the various limiting phenomena and properties at both the micro and macroscales, where the local conditions and environment around the reaction site impact reactivity. In addition, we will take a deeper examination into the transport processes within the ionomer itself. While an ideal ionomer in many different technologies is a solid-state single-ion conductor, this is rarely seen in practice where other ions may exist as contaminants, additives, or due to electrolyte solutions. We will examine these various tradeoffs and their impacts on properties, water transport, and eventually overall cell performance.

166b Electrode Transport Diagnostics and Design Advances for Proton Exchange Membrane Water Electrolyzers

Shawn Litster, Kara Ferner, Fausto Pasmay, Mary Ebbert

To address climate change and environmental pollution, it is crucial we decarbonize the sectors that are both technically and economically challenging to electrify, such as heavy-duty ground transportation, aviation, and industries like ammonia, steel, petroleum refining, synthetic fuels, and cement production. Strategies to reduce the reliance of these sectors on fossil fuels and cutting emissions involves using hydrogen with low carbon intensity. Proton exchange membrane water electrolyzers (PEMWEs) play a vital role in this transition due to their high efficiency, rapid response time, and capability to electrochemically compress hydrogen. Despite encouraging progress, significant improvements in catalyst activity and utilization are necessary to lower loadings to levels that meet ambitious cost targets.

One critical aspect of enhancing catalyst performance for achieving high currents at efficient low voltages is minimizing overpotential losses. These losses result from electron, proton, and water transport to the oxygen evolution catalyst. Additionally, there are secondary current-induced loss mechanisms, including membrane and electrode dry-out and ionic contaminant polarization.

In this presentation, we will delve into the relevant transport mechanisms and the fundamental limits of reducing these overpotentials through modeling. We have evaluated the transport properties of electrodes that are challenging to directly probe due to their micrometer and sub-micrometer length scales using ultra-high-resolution three-dimensional imaging with X-ray computed tomography (nano-CT) and plasma-focused ion beam cross-sectioning with scanning electron microscopy (pFIB-SEM) for image-based simulations. Subsequently, we will briefly review existing methods for experimentally diagnosing transport limitations, as well as highlight new methods currently under development in our laboratory. These new diagnostics include a galvanostatic intermittent titration technique (GITT) for electrolyzers and a limiting-current diagnostic for PEMWEs. The presentation will outline fresh insights into the significance of two-phase transport, electrode and membrane dry-out, all enabled by these diagnostics and modeling. We will conclude by presenting examples of emerging electrode concepts being developed in our laboratory and with collaborators in addressing the transport barriers.

166c Challenges in PEM Electrolyzer Meas

Fan Yang, Qiang Sun, Connor Buek, Irene Mahoney, Cortney Mittelsteadt

Proton Exchange Membrane (PEM) water electrolysis stands out as a leading method for green hydrogen production due to its exceptional efficiency and ability to operate under high differential pressures. At the heart of a PEM electrolyzer lies the Membrane Electrode Assembly (MEA), where water molecules undergo decomposition into hydrogen and oxygen. This presentation will delve into the challenges and future pathways concerning MEA materials, operational aspects, diagnostics, maintenance, and ultimate recycling.

An MEA within an electrolyzer comprises platinum group metal (PGM) catalysts, perfluorosulfonic acid (PFSA) membranes, and additional gasketing materials. The catalyst constitutes a significant portion of the PEM electrolyzer stack's cost. Conventionally, Ir oxide-based catalysts have dominated the PEM electrolyzer industry, but due to their high cost, efforts are underway to reduce catalyst loading and develop more active alternatives. Similarly, the membrane plays a crucial role in the MEA, with various thicknesses such as 5mil and 7mil being utilized. However, thicker membranes contribute to higher resistance, leading to efficiency losses. Hence, reducing membrane thickness has become pivotal for improving PEM electrolyzer performance. Yet, thinner membranes bring challenges like increased hydrogen gas crossover, necessitating efficient gas recombination technologies. Additionally, the high differential pressure operation requires reinforcement layers within the membrane, akin to fuel cell technologies.

The unique characteristics of MEA components demand ultra-pure water inputs, although ensuring this purity can be challenging. Contaminants from plumbing systems and incomplete removal by deionization systems can lead to ion leaching, which in turn poisons the catalysts. Establishing a universal method for identifying and rectifying such issues is imperative for maintaining MEA performance.

MEA degradation over time is inevitable, with an expected lifespan of 80,000 to 100,000 hours. While large-scale PEM electrolyzer applications are still in their infancy, MEA recycling is already a consideration. Despite containing valuable materials such as PGM catalysts and PFSA-based membranes, conventional recycling methods like calcination are suboptimal due to the presence of PFSA membranes. Burning MEAs not only wastes expensive membranes but also poses environmental hazards due to carbon and fluoride emissions. Research efforts are thus focused on catalyst recycling and finding alternative uses for PFSA materials, ensuring a sustainable approach to MEA disposal.

166d A Design of Gas Diffusion Media for Polymer Electrolyte Membrane Fuel Cells

Iryna Zenyuk

Polymer exchange membrane (PEM) fuel cells are promising candidates to decarbonize transportation sector. They use hydrogen to produce electricity and water. The components must be optimized for either high efficiency or high current density operation. Micro-porous layers (MPL) that contact catalyst layer serve as layers that conduct electricity and heat, remove water and enable oxygen delivery to the active sites. The MPL is made of carbon black powder bound by hydrophobic agent. Commercially available MPLs are either impregnating the gas diffusion layer or are free-standing. Some feature cracks and inhomogeneous surfaces that are not favorable for thin (~10 micrometers) membranes that are currently being used in PEM fuel cells community. In this presentation I will summarize the effort into optimizing commercial MPLs and also designing novel MPLs using Teflon AF as a binder. We further try to understand the impact of MPL cracks on the PEM fuel cells durability, a topic that is of importance but has not been studied thoroughly in the field. We conclude that there is still room to advance MPLs to enable broader deployment of PEM fuel cells.

166e Advances in PEFC Short-Stack Diagnostics: Understanding MEA Durability Enhancements for Heavy Duty Fuel Cell Applications

Timothy Van Cleve, Rafiq Ahmed, K.C. Neyerlin

The transportation sector currently accounts for a large portion of global greenhouse gas and particulate emissions, contributing to many adverse effects. Decarbonization efforts will rely on the rapid development and deployment of electric vehicles to replace existing fleets. Hydrogen fuel cells have received considerable interest as a cleaner replacement for diesel engines for long distance transportation. To outcompete diesel engines, fuel cell systems must operate with high fuel efficiency and durability to lower the total cost of ownership to a competitive level. With these ambitious targets, we need to rapidly integrate promising materials in membrane electrode assemblies (MEAs) to assess and improve their performance and durability.

The development of high-performance and durable MEAs is challenging due to the difficultly of optimizing many interrelated processing and fabrication variables (e.g. ink formulation and mixing, deposition method, solvent ratio, hot-pressing, conditioning, etc) that can significantly impact overall performance and durability. Unfortunately, this process is highly empirical and therefore requires substantial time and resources when incorporating a novel component. There's a clear need to accelerate the ability to assess MEA durability and identify key design parameters to advance the development of hydrogen fuel cell technology.

To address the limitations in testing capacity, we utilize short stacks to rapidly screen MEAs up to 10 times faster than single cell testing. We have developed several testing protocols for short stacks that approximate the potential cycling in catalyst, membrane, and MEA accelerated stress tests. The performance losses of individual cells can be monitored throughout testing in rainbow-stack and repeat MEA configurations. Levering our experience in electrochemical diagnostics development, we have recently developed galvanostatic techniques to monitor critical MEA properties (like hydrogen crossover, electrochemical surface area, and ohmic resistance) of individual cells in the stack. Coupling electrochemical characterization with high throughput performance screening, our aim is to fast-track the discovery of structure-performance relationships essential to durable fuel cell development. This presentation will highlight these new diagnostic capabilities and our recent results evaluating advanced MEAs developed within the Million Mile Fuel Cell Truck consortium.

166f Novel Seawater Electrolyzer Design with High-Rate Passive Water Transport

Mohammed Al-Murisi, Masoomeh Ghasemi, Xiaoxuan Yang, Yasmin Teixeira Bimbatti, Jeffrey McCutcheon, Paul Kohl, Sirivatch Shimpalee, William Mustain

As the world pushes decarbonization efforts toward achieving Net Zero carbon emissions, green hydrogen has received an enormous amount of attention as a promising energy carrier. However, for each kilogram of hydrogen to be produced, 9 kg of water needs to be consumed. Traditionally, fresh water is used as the inlet source for electrolysis, but the very large volume of clean water that will be required for H_2 production in the coming decades has the potential to worsen the shortage of fresh water in many parts of the world.

To avoid the competition between energy, agriculture and residential water uses, the use of seawater as a water source has been extensively discussed in recent years. Unfortunately, direct seawater electrolysis is plagued by several challenges including the formation of competitive chloride by-products at the anode, electrode corrosion, and precipitation of divalent cations at the cathode limiting its performance and lifespan. To avoid this, traditional systems first desalinate the seawater by reverse osmosis (RO) before being utilized for green hydrogen production. But the energy intensity of RO coupled with membrane fouling and increased system complexity makes it preferable to operate without the RO step. Therefore, modern solutions often focus on designing corrosion-resistance electrodes, these technologies are still far away from practical implementation, and limiting the operating voltage (which increases capital cost and system size).

Therefore, this work focused on overcoming the issues associated with direct seawater electrolysis using a novel design that is called the osmosis-driven electrolyzer cell (ODEC). This reactor places concentrated electrolyte solutions and simulated seawater on either side of membranes that allow for water transport into the electrolyte solutions without significant ion transport. The electrolyte solution accepts the water from the seawater and that solution is used to drive electrolysis and H_2 generation. This talk will give an insight into multiple cell designs around the ODEC concept, with a discussion of strengths and weaknesses. We systematically manipulated membrane properties, cell geometry and cell operating conditions to improve device efficiency and performance. The experimental results are supported by computational fluid dynamics modeling, providing further insight into the cell behavior. Lastly, data will be shown where electrolyzers are durably run for 100's of hours.

238a Tuning Transport and Reaction across Length Scales for Next Generation Electrochemical Devices

Jacob Spendelow

Electrochemical systems such as fuel cells, electrolyzers, and reversible fuel cells require complex multiphase transport and reaction at poorly defined interfaces. This talk will describe new approaches to design both catalysts and electrodes for faster transport and reaction. We have recently demonstrated that the use of intermetallic nanoparticle catalysts such as $L1_0$-PtCo can enable significant improvements in long-term performance due to improved stabilization of base metal in the ordered lattice, while use of carbon supports with controlled pore size distribution provides further advantages by protecting metal nanoparticles from degradation or poisoning by ionomer. Recent results and future directions for this catalyst research will be discussed. While kinetic performance and durability of these catalysts is promising, performance at high power densities continues to be limited by slow O_2 transport. Several factors, including slow diffusion through ionomer films, Knudsen diffusion through micropores, and pore blockage by liquid H_2O contribute to these transport limitations. Design of unconventional electrode structures that can minimize these limitations is needed to enable full utilization of cathode catalysts. New hierarchical electrode structures capable of reducing these transport losses will be presented.

238b Development of Ru-Based Electrocatalysts for Hydrogen Oxidation Reaction in Alkaline Media

Minhua Shao

The reaction kinetics of hydrogen oxidation reaction (HOR) in alkaline solutions is much slower than that in acidic media for noble metals like Pt and Pd. At low pH, Ru is not stable even at low potentials, which prevents it from being used as the main components in proton exchange membrane fuel cells. However, it can be much more stable at high pH. In this talk, I will present some recent progress in my lab developing Ru-based electrocatalysts and their fuel cell performance.

238c Overcoming Challenges in Ion-Pair Membrane Technology: Advancements in Fuel Cells and Electrochemical Hydrogen Pumps

Manjeet Chhetri, Youngkwang Kim, Sandip Maurya, Yu Seung Kim

The Membrane Electrode Assembly (MEA) stands as the pivotal element in electrochemical devices such as fuel cells and electrochemical hydrogen pumps (EHP). Notably, ion-pair High-Temperature Proton Exchange Membrane Fuel Cells (HT-PEMFCs) with reduced phosphoric acid concentrations within the MEA have been innovatively designed, allowing for the integration of diverse ion-conducting binders, known as ionomers, into both the cathode and anode electrodes **[1]**. This study delineates initial distinctions in MEA configurations among different device types (HT-Fuel cell and EHP, conducting a comprehensive analysis of performance disparities between traditional polybenzimidazole-based systems and various ion-pair proton exchange membranes **[2]**.

Subsequently, a detailed exploration unfolds, focusing on the utilization of ion-pair membranes and different ionomer types for the three aforementioned devices. This comprehensive examination offers a unified comparative analysis encompassing ionomer type, membrane thickness, catalyst:ionomer ratio, and phosphoric acid doping levels, particularly in the context of performance at intermediate temperatures (160°C). Conclusions are drawn based on results derived from both half-cell and single-cell evaluations. Finally, overarching guidelines are presented to inform the formulation of high-performance MEA configurations tailored for ion-pair HT-PEMFCs and EHPs.

References:

[1] Lim, K. H.; Lee, A. S.; Atanasov, V.; Kerres, J.; Park, E. J.; Adhikari, S.; Maurya, S.; Manriquez, L. D.; Jung, J.; Fujimoto, C.; Matanovic, I.; Jankovic, J.; Hu, Z.; Jia, H.; Kim, Y. S. Protonated phosphonic acid electrodes for high power heavy-duty vehicle fuel cells, *Nature Energy*, 7, 248 (2022).

[2] Lim, K, H; Matanovic, I; Maurya, S; Kim Y; De Castro E. S; Jang, J-H; Park, H; Kim, Y.S. High Temperature Polymer Electrolyte Membrane Fuel Cells with High Phosphoric Acid Retention. *ACS Energy Lett.*, 8, 1, 529–536 (2023).

238d Anchoring Pt Catalyst with ZrO_2 on Carbon Black for Enhanced ORR Durability

Gan Xu, Sajid Hussain, Jingxin Zhang, Nagappan Ramaswamy, Swami Kumaraguru, Yangchuan Xing

Carbon black-supported platinum catalysts (Pt/C) remain the mainstream oxygen reduction reaction (ORR) catalysts for proton exchange membrane fuel cells (PEMFCs). However, the high cost of Pt and its degradation during operation hinder the practical application of PEMFCs. To solve this problem, zirconia (ZrO_2) has been considered for anchoring platinum nanoparticles (NPs) onto carbon surfaces due to its excellent acid/base resistance, aiming to reduce Pt loading and enhance the durability of Pt/C catalysts. In this project, a facile deposition of ultra-thin ZrO_2 layers onto the surface of Vulcan XC72 carbon black was achieved through the condensed layer deposition technique. Subsequent high-temperature annealing resulted in uniformly distributed ZrO_2 nanoparticles (~2.0 nm), as confirmed by transmission electron microscopy (TEM). The deposition of Pt NPs (~3.5 nm) was accomplished by ethylene glycol reduction of potassium tetrachloroplatinate under heating conditions. The final composition of ZrO_2-anchored Pt/C catalyst ($Pt/ZrO_2/C$) was determined to be 22.9 wt% Pt, 4.9 wt% ZrO_2, and 72.2 wt% C through thermogravimetric analysis. The catalysts were characterized by high-resolution TEM, energy dispersive X-ray spectroscopy, and X-ray photoelectron spectroscopy, confirming the presence, formation and distribution of Pt and Zr. Corresponding ORR and catalyst durability tests demonstrated excellent activity and stability of $Pt/ZrO_2/C$ in acid media. After 10,000 cycles between 0.6 and 1.1 V_{RHE} in 0.1 M perchloric acid, there was only ca. 10 mV shift in the half-wave potential, with no significant change in the electrochemical surface area. It was demonstrated that ZrO_2 can be an anchoring metal oxide to enhance the catalyst durability.

238e Highly Durable Hydrogen Fuel Cell Cathodes for Heavy Duty Transportation Applications

Marwa Atwa, Fritz Prinz

The demand for clean and sustainable energy sources is increasing globally, with hydrogen emerging as a prominent energy carrier. Polymer exchange membrane fuel cells (PEMFCs) offer a promising solution, converting the chemical energy of hydrogen into electricity with zero emission for many industrial applications, including automotive applications. While hydrogen-fueled PEMFCs provide higher efficiency than internal combustion engines, a substantial enhancement in durability while maintaining high efficiency and performance is needed compared to the current state of the technology for light-duty vehicles. To meet the heavy-duty application demand, significantly longer lifetimes (4-5 times improvement in durability) with high efficiency are the main targets.

In this presentation, we present the nanoengineering development of highly durable and high-performance efficient cathodes for PEMFCs, utilizing mesoporous carbons as supports and platinum as the electrocatalyst. Through our study, we have established novel cathodes that allow high catalytic activity, performance, durability, and carbon corrosion resistance—a very rare combination to be found in one material. To obtain this material, a range of mesoporous carbon supports with different porosity and graphitization degrees have been tested for performance and durability in a single-cell PEMFC device. Screening of these mesoporous carbon supports enables us to reveal the main characteristics of a carbon support that can lead to a robust PEMFCs cathode that can mitigate platinum dissolution and performance losses after extended several thousands of accelerated durability test while being itself corrosion resistive.

238f High-Performance PGM Catalysts and Meas for Heavy-Duty Fuel Cells

Gang Wu

Proton exchange membrane fuel cells (PEMFCs) are a promising zero-emission power source for heavy-duty vehicles (HDVs). However, long-term durability of up to 25,000 h is challenging because current carbon support, catalyst, membrane, and ionomer developed for traditional light-duty vehicles cannot meet the stringent requirement. Therefore, understanding catalyst degradation mechanisms under the HDV condition is crucial for rationally designing highly active and durable platinum group metal (PGM) catalysts for high-performance membrane electrode assemblies (MEAs). Herein, we report a PGM catalyst consisting of platinum nanoparticles with a high content (40 wt %) on atomic-metal-site (e.g., MnN_4)-rich carbon support. MEAs with the Pt (40 wt %)/Mn–N–C cathode catalyst achieved significantly enhanced performance and durability, generating 1.41 A cm^{-2} at 0.7 V under HDV conditions (0.25 mg_{Pt} cm^{-2} and 250 kPa_{abs} pressure) and retaining 1.20 A cm^{-2} after an extended and accelerated stress test up to 150,000 voltage cycles. Electron microscopy studies indicate that most fine Pt nanoparticles are retained on or/and in the carbon support covered with the ionomer throughout the catalyst layer at the end of life. During the long-term stability test, the observed electrochemical active surface area reduction and performance loss primarily result from Pt depletion in the catalyst layer due to Pt dissolution and redeposition at the interface of the cathode and membrane. The first-principle density functional theory calculations further reveal a support entrapment effect of the Mn–N–C, in which the MnN_4 site can specifically adsorb the Pt atom and further retard the Pt dissolution and migration, therefore enhancing long-term MEA durability.

362a Linking Structure to Function at Electrochemical Interfaces: Li-Ion and Beyond

Lauren Marbella

Despite the fact that the solid electrolyte interphase (SEI) on Li metal was described 45 years ago, it is still the only aspect of the battery that has ambiguity in function. As a community, we have struggled to establish structure-property-performance relationships for the SEI because it is a nanoscale composite that contains chemical compounds whose properties deviate from their bulk counterparts. In this talk, I will describe how we have used nuclear magnetic resonance (NMR) spectroscopy to characterize the structure and dynamics of interfacial phenomena in Li-ion and beyond Li-ion batteries and correlate these features with battery performance. In particular, I will focus on the use of NMR to quantify the source of Li inventory loss, the mechanism of transition metal dissolution, structural evolution at the electrode/electrolyte interface, and the function of the SEI. Insight from these methods allow us to determine the precise mechanisms of failure that arise inside of functional devices as well as develop new approaches to mitigate performance decline.

362b A Versatile Optimization Framework for Porous Electrode Design: Coupling a Genetic Algorithm and a Pore Network Model

Maxime Van der Heijden, Gabor Szendrei, Victor de Haas, Antoni Forner-Cuenca

Porous electrodes are performance- and cost-defining components in modern electrochemical systems as they determine the hydraulic resistance, facilitate mass transport, conduct electrons and heat, and provide surfaces for electrochemical reactions [1]. Thus, electrode engineering is an effective approach to improve cost competitiveness by increasing power density. In convection-enhanced technologies, currently used porous electrodes are fibrous substrates developed for low-temperature fuel cells, but their microstructure and surface chemistry limit the performance of emerging electrochemical systems. Microstructure-informed multiphysics simulations can be leveraged to aid the theoretical design of advanced electrode architectures [2]. However, they have only recently been deployed for the bottom-up design of electrode microstructures [3]. The combination of microstructure-informed multiphysics with evolutionary algorithms could accelerate progress in the optimization of porous electrodes for a given application. In this work, we combine three-dimensional simulations with a genetic algorithm for the bottom-up design of porous electrodes for redox flow batteries.

In the first part of the talk, I will describe a methodology to couple an experimentally validated pore network modeling framework that is microstructure-informed and electrolyte-agnostic [4], with an evolutionary algorithm [5]. This genetic algorithm is used to optimize electrode microstructures by evolving the structure driven by a fitness function that minimizes pumping power requirements and maximizes electrochemical power output, where the optimization only relies on the electrolyte chemistry and initial electrode and flow field geometries as inputs. The analyzed proof-of-concept employs a flow-through cubic lattice structure with fixed pore positions and shows significant improvement of the fitness function over 1000 generations. The fitness improved by 75% driven by a reduction in the pumping requirements by 73% and an enhanced electrochemical performance of 42%. The evolutionary design resulted in a bimodal pore size distribution containing longitudinal electrolyte flow pathways of large pores and an increased surface area at the membrane-electrode interface.

In the second part, I will discuss our latest progress including the introduction of geometrical versatility by adding a pore merging and splitting function, the impact of various optimization parameters, geometrical definitions, and objective functions, and the incorporation of electrode structures and flow fields with well-defined geometries [6]. Moreover, I will show the need for optimizing electrodes for specific reactor architectures and operating conditions to design next-generation electrodes by analyzing the optimization for initial starting geometries with diverse flow field designs (flow-through and interdigitated), morphologies (cubic and a tomography-extracted commercial electrodes), and redox chemistries (VO^{2+}/VO_2^{+} and TEMPO/TEMPO^{+}). The presented genetic algorithm offers potential for the predictive design of high-performance electrode microstructures for a broad range of operating conditions, electrolyte chemistries, reactor designs, and electrochemical technologies. While applied to flow batteries in this study, this methodology can be leveraged to advance electrode microstructures in other electrochemical systems by adapting the relevant physics.

Acknowledgments

The authors gratefully acknowledge the Dutch Research Council (NWO) through the Talent Research

Program Veni (17324) for financial support.

References

[1] M. van der Heijden, A. Forner-Cuenca, *Encyclopedia of Energy Storage*, 480-499 (2022)

[2] B. Chakrabarti *et al., Sustainable Energy and Fuels,* **4**, 5433-5468 (2022)

[3] V. Beck *et al., J. Power Sources,* **512**, 230453 (2021)

[4] M. van der Heijden & R. van Gorp *et al.*, *J. Electrochem. Soc.*, **169** 040505 (2022)

[5] R. van Gorp & M. van der Heijden *et al.*, *Chem. Eng. J.*, 139947 (2022)

[6] M. van der Heijden *et al.*, *Digital Discovery*, Accepted manuscript (2024)

362c Electrolyte Engineering in Battery Technology

Ge Li

Electrolyte engineering is one of the most promising approaches that can revolutionize battery technology in terms of energy, power, safety, and extreme condition operations. Alkali metal batteries, by using the metallic anode (e.g., Li, Zn) can significantly increase the energy density of the battery due to their high theoretical specific capacity, low density, and low electrochemical potential. However, large-scale deployment of metal anodes in practical batteries still requires a scientific breakthrough in solving major challenges of dendrite growth and low plating/stripping efficiency of metal anodes. To address the issues related to metallic anodes in alkali metal batteries, several effective strategies have been developed to stabilize metal anodes, including liquid electrolytes, surface coating, solid-state electrolytes, and porous current collectors. In this presentation, we will present our recent findings on the development of suitable electrolyte systems for next-generation alkali metal batteries.

362e Synergistic Coupling of Anionic Oxygen Redox with Selenium for Stable High-Voltage Sodium Layered Oxide Cathodes

Neha Bothra, Jagjit Nanda, Xueli Zheng, Michal Bajdich

The increasing need for accessible and sustainable energy storage solutions has brought sodium-ion (Na-ion) cathode materials[1,2] to light as possible contenders for the upcoming generation of battery cathodes. Unfortunately, Na-ion layered cathodes' irreversible oxygen redox reaction has made it difficult for them to be widely used, resulting in voltage deterioration and a shorter lifespan. In this work, we conduct a thorough investigation of the $Na_{0.6}Li_{0.2}Mn_{0.8}O_2$ classic model Na-ion cathode material, demonstrating a notable spatial heterogeneity in its redox couple progression from the bulk to the surface. When oxygen is released, cationic valence states stratify from the surface to the bulk, causing surface cations to become 'activated' and engage in electrochemical redox while inhibiting anionic redox at the same time. This process progressively spreads to the particle's bulk phase.

We design Se-oxidative doped surfaces to counteract this undesirable transfer of redox coupling. Significantly reducing manganese redox activity, this revised cathode exhibits steady voltage behavior and maintains reversible oxygen redox even after 100 cycles. While the pristine $Na_{0.6}Li_{0.2}Mn_{0.8}O_2$ cathode decreases from 58% to 14%, the anionic redox capacity contribution of the Se-doped $Na_{0.6}Li_{0.2}Mn_{0.8}O_2$ cathode stays high at 45% from an initial 62%. The comprehensive computational electronic analyses have offered us valuable insights into the enhanced stability observed in this context. These results highlight the possibility of using sophisticated synergistic coupling design to lessen the problem of irreversible anionic redox-induced voltage deterioration.

362f Design of Electrolyte Molecules for Lithium Batteries

Nan Yao, Yi Yang, Yu-Chen Gao, Legeng Yu, Chong Yan, Xiang Chen, Qiang Zhang

Lithium batteries (LBs) have closely integrated into sustainable energy systems by powering electronic devices and facilitating grid-scale energy storage. Meanwhile, the expanding requirements and the diversification of application scenarios appeal to next-generation LBs with superior performance, including fast-charging capability, all-climate adaptability, high energy density, and long lifespan. This puts forward increasingly high demands on battery materials to shape future advanced LBs.

A typical LB is composed of two electrodes and a liquid electrolyte. The liquid electrolyte functions as the "blood" of a battery, connecting the two electrodes and serving as the medium for ion transport. Its properties play a direct role in determining the practical performance of the battery as opposed to the theoretical upper limit set by the electrodes. Therefore, electrolyte design is perceived as the primary solution to satisfying the requirements of multifarious applications and constructing high-performance LBs.

However, electrolytes are complex systems which consist of various components. Researchers have adopted conventional trial-and-error methods to study thousands of molecules over the past years, but only dozens of molecules haven been found to be working in LB electrolytes. Considering the vast molecular space, this time-expense and low-efficiency approach renders it impractical to meet the urgent and diverse demands for electrolytes in the future, which presents both a significant opportunity and a formidable challenge for the electrolyte design.

Here, we integrate multiscale simulations and experimental methods to design electrolytes, including elucidating the structure–property relationships of electrolytes, evaluating electrolyte properties by computational methods, and designing electrolytes molecules according to target needs. First, the stable cycling of a battery is the prerequisite for satisfying all other requirements, so the electrolyte stability is focused on primarily. Models of the electrolyte solvation structures were constructed to investigate the dependence of the electrolyte stability on electrolyte structures. By investigating on around 1400 systems, it is revealed that the electrolyte oxidative and reductive stability increases and decreases, respectively, as long as the Li^+–molecule solvation structures are formed. This emphasizes the significance of considering the solvation structures instead of stand-alone molecules in evaluating the electrolyte stability. Second, a high-throughput electrolyte calculation (HTEC) software is self-developed. HTEC consists of 8-round density functional theory calculations and molecular dynamics simulations, which can achieve a convenient and efficient evaluation of over 20 kinds of electrolyte properties including stability, viscosity, conductivity, etc. An electrolyte database contains over 200,000 entries have been constructed for now, which is the biggest one around the world to the best our knowledge. Third, electrolyte design based on different application requirements is carried out by target-orientated screening of molecules. Taking the design of fast-charging LB electrolytes at high temperatures as an example, the thermal stability, chemical stability, binding strength to Li^+, viscosity, boiling point, etc., which play significant role in determining the fast-charging performance of electrolytes at high temperatures, are considered. Two promising molecules in carboxylic ester structures are discovered by the stepwise screening according to property criteria. These two novel molecules render superior battery performance compared with the commercialized LB electrolytes (ethylene carbonate/dimethyl carbonate/lithium hexafluorophosphate-based).

To conclude, our work establishes quantitative structure–property relationships and leveraging the high-throughput advantages of computational methods to expand the molecular space under investigation, thereby expediting the design of promising molecules through a highly efficient data-driven approach. It not only has demonstrated success in the electrolyte design but also holds great potential in the design of other functional materials.

362g Experimental and Computational Characterization of Cathodes for LiS Batteries

Andrea Querio, Daniele Versaci, Gianluca Boccardo, Daniele Marchisio, Silvia Bodoardo

Nowadays, lithium-ion batteries (LIBs) are playing a crucial role in the field of sustainable renewable energy sources due to their large versatility which allows them to be used in different applications such as energy storage systems, electric vehicles and electric portable devices. However, considering the further increase in global demand of batteries and electric performance, it is important to underline two key factors which are limiting the future employment of LIBs in the market: the supply of critical raw materials (CRMs) such as transition metals like nickel, manganese and cobalt, and the practical specific capacity of commercial LIBs. These limitations are leading to research on new energy storage systems, such as lithium-sulfur (LiS) and lithium-air (LiO_2) batteries, which are growing exponentially as future alternative of LIBs. Both systems have great advantages in respect to lithium-ion, such as high energy density, low cost, high availability of active material and they are more sustainable. In particular, LiS batteries seem to be more suitable to be commercialized in the coming years.

However, LiS systems suffer from two main limitations: the shuttling effect and the precipitation of active species, both leading to capacity fade and cell degradation. The first one consists in the migration of long-chain polysulfides from the cathode to the anode where they can react to form shorter polysulfides or lithium sulfides, which precipitate directly on the anode surface, involving both loss of active materials and anode corrosion, contributing to self-discharge. On the other side, the final products of charge and discharge processes (sulfur and lithium sulfide) are insoluble and precipitate, driving to loss of active species and deterioration of cathode structure due to the low electrochemical reversibility of these reactions. Both phenomena can be contained working on the cathode formulation by changing different features: the physico-chemical nature of the carbonaceous matrix used and/or the addition of different catalyst materials such as metal sulfides, metal oxides etc. However, the optimization of operating conditions generally requires several time-consuming and resource-demanding experiments. For this purpose, the employment of a computational model, able to describe the discharge/charge behaviour of the cell, and the effect of different operating conditions on the electrochemical performances, can be a suitable option to predict the behaviour of the system.

The present work has two main objectives: the first one is the experimental characterization of cathodes for LiS batteries, studying different compositions and formulations in order to optimize the electrochemical performances of the cells; the second one is the development of a physics-based model that can accurately describe all the elements present in a battery and all the phenomena that occur during charge and discharge processes at the pore-scale level. The model developed is a continuum physics-based model that describes the battery at the pore-scale, focusing on a realistic description of the porous cathode structure, which is represented by a separate domain, while the separator is modelled as a continuum domain using the Bruggeman correlation for the diffusion coefficients, and the anode consist of a lithium foil. The reaction chain is modelled with two chemical reactions, *i.e.*, the dissolution of sulfur and the precipitation of the lithium sulfide, and five electrochemical reactions that characterize the reduction of the long-term chain polysulfide to sulfur anion. The two chemical reactions are modelled considering the difference between the concentration of the ions involved and the solubility product of the solid component as driving force; instead, the electrochemical reaction rate is evaluated from the current intensity generated from the reaction, which in turn is determined employing the Butler-Volmer equation. The model can predict the discharge behaviour of the cells and can describe the evolution during the discharge process of species concentrations, which can provide further insights on the shuttling phenomena and the sulfide precipitation, allowing to determine the effect of different conditions on these phenomena and on the electrochemical behaviour of whole system.

The preparation of the active material consists of different steps. First, sulfur and carbon are mixed together with a polyvinylidene difluoride (PVdF) binder using N-methyl-2-pyrrolidone (NMP) as solvent. In this step, three different features can be investigated: the mass ratio between sulfur, carbon and binder, the carbonaceous type employed and the technique used to mix all the materials. In our work, we started with a mass ratio of 60:30:10 between the three main components, progressively changing the amount of sulfur in order to increase the mass loading of the cathode to obtain a larger specific energy. The second feature investigated was the nature of the carbonaceous matrix, and two different types of conductive carbon, a mesoporous carbon and a nanofiber carbon, were considered. These two materials differ in morphology, porosity and specific surface which influence both the precipitation of lithium sulfides and the migration of polysulfides. Nanofibers show improvements for both effects, favoring the formation of small lithium sulfide particles which can be easily dissolved (avoiding the formation of lithium sulfide layer which are more insoluble) and, at the same time, enhancing the retention of sulfur inside the electrode. The third feature studied was the techniques employed to mix together the materials: a classical mechanical mixing performed by ball milling and a more peculiar method based on melt-infusion, which consists of melting the sulfur at high temperature enhancing its diffusion inside the pores of the conductive carbon. The melt-infusion method favours the retention of sulfur inside the electrode, decreasing the shuttling effect phenomena. After this step, the active material was cast on a carbon coated aluminium foil current collector using a Doctor Blade to control the thickness of the electrode, which was changed within 200 μm to 300 μm.

To assemble coin cells (2032), 15 mm diameter disks were cut from the cast material, to be used as cathode, while metallic lithium disks were used as anode, and 1 M LiTFSI in 1,3-dioxolane:dimethoxy ethane 1:1 (DOL:DME, 1:1) with 0.25 M $LiNO_3$ was used as electrolyte. The electrodes were dried in a vacuum oven before assembling inside an argon-filled glove box, and a commercial separator (Celgard 2500) was also used. The coin cells were used to perform electrochemical testing to characterize the cathodes produced and their electrochemical behaviour. Different types of tests were carried out: galvanostatic cycling and rate capability tests were employed to evaluate the cycle life of the cells and their behaviour at different current intensity, while cyclic voltammetry were used to evaluate thermodynamics and kinetics information about the redox processes. Other experiments can be employed: intermittent current interruption (ICI) was used to evaluate the evolution in time of the internal resistance and diffusion coefficient of the lithium ions in the electrode, potentiostatic intermittent titration technique is useful to obtain further information on the nucleation and growth of Li_2S crystals during discharge, field emission scanning electron microscope (FESEM) was used to characterize cathode morphology in order to represent more realistically the electrode structure. These tests can be employed to validate the model, comparing the results with the computational predictions, but also to inform the model and to describe more accurately the cell and the phenomena occurring during charge/discharge processes.

This work was funded by the Stellantis-CRF Polito framework agreement 2022-2026.

362h Enabling Transition-Metal-Free and Solvent-Free Batteries

Chibueze Amanchukwu

Batteries with high energy densities (energy stored per mass or volume) are required to electrify transportation and the grid. However, current Li-ion batteries are plagued by the presence of critical minerals such as Co, Ni that are geographically limited and increase costs. In addition, the presence of volatile and flammable electrolyte solvents has led to a poor safety profile and significant hazards. Here, we present a new class of solvent-free and transition-metal free dual graphite batteries. This battery chemistry exploits cation intercalation at the anode and anion intercalation at the cathode with graphite serving as both electrodes. We probe the changes to the electrochemical behavior as a function of electrolyte composition. Our work illustrates the future of battery design that does not compromise safety or cost for energy density and can enable long duration energy storage.

703a Mass Transport and Interfaces in CO2 Electrolysis Systems

Feng Jiao

Traditional chemical industry processes often rely on fossil fuels, which inevitably emit substantial quantities of CO_2. Turning CO_2 into valuable chemicals is a potential approach for sustainable chemical production. In this presentation, we will discuss our work on understanding and engineering interfaces in CO_2 electrolysis system for acetate and ethylene production. We will share our initial discovery of acetate formation in CO electroreduction on Cu catalysts. Mechanistic studies suggest that a ketene-like intermediate is the key for the formation of acetate. With further efforts in reactor engineering, we successfully developed an internally coupled purification strategy that significantly enhances acetate concentration and purity in CO electrolysis.[1] This approach employs an alkaline-stable anion exchange membrane with high ethanol permeability and a selective ethanol partial oxidation anode to regulate the CO reduction product stream. Furthermore, we will discuss our recent efforts and technical challenges in scaling up tandem CO_2 electrolysis systems for industrial applications. [2]

References:

1. S. Overa, B. Crandall, B. Shrimant, D. Tian, B. H. Ko, H. Shin, C. Bae and F. Jiao*. Enhancing acetate selectivity by coupling anodic oxidation in carbon monoxide electroreduction. Nature Catalysis 5, 738-745 (2022). 10.1038/s41929-022-00828-w
2. F. Jiao, et al. Tandem CO2 Electrolysis: From Watt to Kilowatt-Scale for Enhanced Acetate and Ethylene Production. Nature Chemical Engineering (in press).

703b Experimental Trends and Theoretical Descriptors for Electrochemical Reduction of Carbon Dioxide to Formate over Sn-Based Bimetallic Catalysts

Xue Han, Binhong Wu, Yao Nian, Qiaowan Chang

The electrochemical carbon dioxide reduction reaction (CO_2RR) using renewable energy sources is a promising solution for mitigating CO_2 emissions. In particular, CO_2RR to formate represents a commercially profitable target. However, a comprehensive understanding of the catalytic mechanisms of Sn-based catalysts under reaction conditions, including the real-time structural evolution of catalysts and the role of all key reaction intermediates in influencing the CO_2RR selectivity, is still lacking. The current study reports a framework to study the selectivity preference of Sn-based bimetallic catalysts using a combination of electrochemical measurements, in-situ characterization, and density functional theory (DFT) calculations. The addition of a second metal (Co, Ni, Ag, Zn, Ga, Bi) was found to play a vital role in affecting the CO_2RR performance. *In-situ* X-ray absorption near edge structure (XANES) measurements revealed a dynamic evolution in the Sn valence state induced by different secondary metals. A multidimensional descriptor involving all the key reaction intermediates was developed to assess formate selectivity using a 2-dimensional volcano plot. This research offers an effective framework for understanding CO_2RR catalytic selectivity by considering both the real-time structural evolution of catalysts and all the key intermediates involved.

703c Stories about Single Atom Catalysis

Huabin Zhang

Increasing demand for sustainable energy has accelerated research on various renewable technologies. Developing catalytic energy conversion technologies for replacing the traditional energy source is highly expected to resolve the fossil fuel exhaustion and related environmental problems. Exploring stable and high-efficiency catalysts is of vital importance for the promotion of these technologies. Single-atom catalysts (SACs), containing single metal atoms anchored on supports, represent the utmost utilization of metallic catalysts and thus maximize the usage efficiency of metal atom.[1] However, with the decreasing of particle size, the surface free energy increases obviously, and tends to aggregate into clusters or particles. Selection of an appropriate support is necessary to interact with isolated atoms strongly, and thus prevents the movement and aggregation of isolated atoms, creating stable, finely dispersed active sites. Furthermore, with uniform single-atom dispersion and well-defined configuration, SACs afford great space for optimizing high selectivity and activity.

Our research interest focuses on the development of single-atom catalysts with the particular configuration for sustainable energy conversion, including photocatalysis, electrocatalysis, and thermal catalysis.[2] His research also extends to the operando investigation for monitoring the structural evolution of the reactive centers, as well as the mutual interaction between the reactive center and reactant in the catalytic process. The widespread adoption of SACs in diverse catalytic reactions will be comprehensively introduced.[3, 4] By presenting these advances and addressing some future challenges with potential solutions related to the integral development of catalysis over SACs, we expect to shed some light on the forthcoming research of SACs for catalytic energy conversion.

Reference:

1. Zhang H, Liu G, Shi L, Ye J. Single-atom catalysts: emerging multifunctional materials in heterogeneous catalysis. *Adv. Energy Mater.* 2018, **8**, 1701343.
2. Feng C, Wu ZP, Huang KW, Ye J, Zhang H. Surface Modification of 2D Photocatalysts for Solar Energy Conversion. *Adv. Mater.*,2022, **34**, 2200180.
3. Xue Z-H, Luan D, Zhang H, Lou XWD. Single-atom catalysts for photocatalytic energy conversion. *Joule*, 2022, **6**, 92-133
4. Duan L, *et al.* A molecular ruthenium catalyst with water-oxidation activity comparable to that of photosystem II. *Nat. Chem.* 2012, **4**, 418-423.

703d Modeling of the Electrochemical Double Layer during CO_2 Electrolysis

Alex J. King, Justin C. Bui, Adam Z. Weber, Alexis T. Bell

An attractive approach to help minimize carbon emissions is the electrochemical conversion of CO_2 into value-added products, such as syngas. However, a critical challenge for this technology is the ability to control the selectivity of the reaction. For CO_2 reduction (CO_2R), the selectivity depends not only on the catalyst identity, but also on the electrolyte environment adjacent to the electrode surface (*i.e.*, the electrochemical double layer). Despite its importance, the structure and composition of the double layer during CO_2 electrolysis remains elusive, largely because of the complex bicarbonate buffer reaction system. Previously published models of the CO_2R double layer often neglect the bicarbonate buffer reactions and do not include model validation. Therefore, there is a need to develop a comprehensive, multiphysics model of the electrochemical double layer during CO_2 electrolysis in order to identify the structure of the double layer and to understand how it impacts CO_2R kinetics.

This talk will present our recent efforts to simulate the electrochemical double layer during CO_2 electrolysis by using a one-dimensional multiphysics continuum model that accounts for the bicarbonate buffer reactions, species diffusion and migration, steric forces, local electrostatic and physical interactions between ions, and the 2nd Wien effect. This model is validated by comparing experimental and simulated capacitance versus potential curves and CO product polarization curves from a Ag catalyst. The model shows how the CO_2R double layer structure (species concentrations and the electrostatic potential) changes as a function of applied potential and electrolyte concentration. Moreover, we explore how the identity of the cation (Cs^+, K^+, Na^+, vs Li^+) impacts the double layer structure and, consequently, CO_2R kinetics. Our model predicts that at CO_2R potentials the Outer Helmholtz Plane (OHP) is dominated by cations and that cations with a small, hydrated radius can pack very tightly at the OHP, thereby increasing the electric field within the Stern layer. We propose that this Stern layer electric field can impact the reaction kinetics by polarizing the neighboring water molecules and facilitating CO_2R. This work offers key insight into the structure of the double layer during CO_2 electrolysis and its role in modulating reaction kinetics, a critical step toward improving control over reaction selectivity.

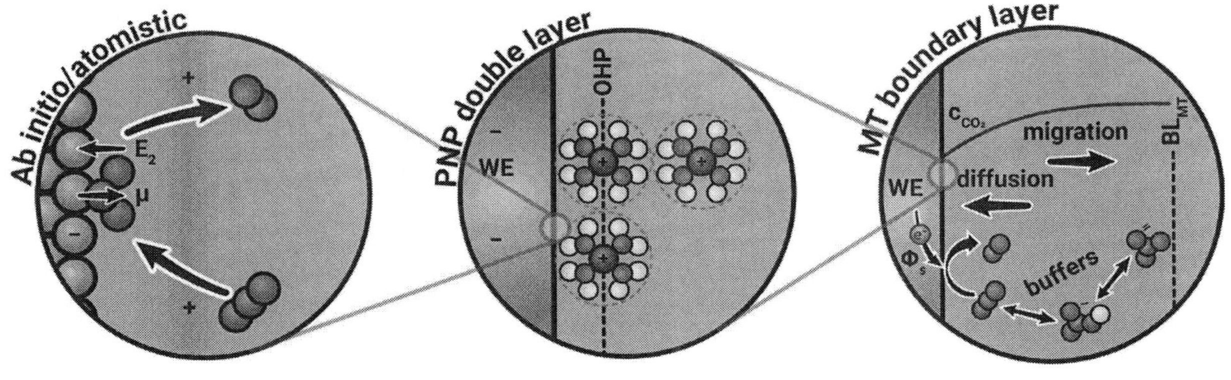

703e Controlling the Electron Transfer Selectivity Via Interfacial Ion Shielding Mechanism

Yiming Mo

Electrosynthesis has recently emerged as a promising tool to synthesize value-added fine chemicals and pharmaceuticals with minimal waste generation and under mild reaction conditions. Steering the direction of electron transfer among multiple electron donors/acceptors is the fundamental task in controlling the chemoselectivity in electrochemical reactions. The precise alignment of redox potentials between electrodes and substrates forms the basis for chemoselective transformations. However, to achieve direct electron transfer to/from a thermodynamically less favored target in presence of more easily reduced/oxidized functionalities, alternative mechanisms are often pursued to circumvent this thermodynamic barrier, such as hydrogen atom transfer, ligand-to-metal charge transfer, and proton-coupled electron transfer. Despite the advancements of these mechanisms, direct electron transfer remains attractive for its inherent simplicity, without the need of additional mediators or pre-functionalization compared to other pathways. However, the lack of other levers for chemoselectivity control beyond matching redox potentials restricts its broader applicability. In this talk, we would like to present a new strategy to steer the electron transfer selectivity. This strategy relies on the heterogeneous ion shielding mechanism by electrostatically adsorbing anions on the anode interface. The formed ion shielding layer imposes mass transfer limitations on the electrode interface that could block undesired molecules to approach the electrode, thus avoiding undesired side-reactions. Thus, electron transfer can selectively proceed between anode and adsorbed anions. To demonstrate the practical value of this new strategy, we developed a photoelectrochemical decarboxylative trifluoromethylation of sensitive (hetero)arenes using trifluoroacetate, an inexpensive yet relatively inert CF_3 source. An ion shielding layer, formed by trifluoroacetate anions electrostatically adsorbed on a positive WO_3 photoanode, prevents undesired electron transfer between substrates and photogenerated holes. The applicability of the developed method was demonstrated with robust photoanode stability (~380 h) and a good substrate scope. To further target practical applications, we designed and demonstrated a scaled-up photoelectrochemical flow cell with 378 cm^2 total electrode area, and achieved 100-gram scale trifluoromethylation of DNA base analogue in the designed flow.

703f Electrocatalysts Design for $2e^-$ Oxygen Reduction Reaction

Xiaolei Wang

Electrochemical production of hydrogen peroxide (H2O2) via 2-electron oxygen reduction reaction (ORR) offers a promising approach for sustainable decentralized production of H2O2. However, ORR can go through either 4-electron pathway to water or 2-electron pathway to H2O2. The need for controllable modulation of ORR pathways, as well as the kinetic challenges, calls for the discovery and optimization of efficient catalysts. Herein, we focus on various design strategies of electrocatalyst towards 2-electron ORR, including (1) develop advanced electrocatalysts with high activity and selectivity for different application media, including alkaline, neutral, and acidic media; (2) elaborate *ex-situ* and *in-situ* characterization of electrocatalyst structure and electrochemical investigation; (3) reveal the underlying factors that contribute to the high performance and provide guidance for future design; (4) build up the gap between lab screening and practical application to facilitate its commercial utilization.

Our research starts with the electrocatalyst design for alkaline electrolyte. We demonstrate a promising carbon-based catalyst, consisting of oxygen-rich hollow mesoporous carbon spheres (HMCS). The as-prepared HMCS exhibit high onset potential (0.82 V) and half-wave potential (0.76 V) as well as excellent H2O2 selectivity (above 95%). The outstanding performance arises from a combination of several aspects, such as porous structure-facilitation of mass transport, large surface area, and proper distribution of oxygen-containing functional groups modification on the surface. Furthermore, the proposed ORR mechanism on HMCS surface reveals that -OH functional groups help promote the first electron transfer process while other oxygen modification facilitates the second electron transfer.

We designed a series of HMCS for the investigation of porosity engineering on the performance. The as-prepared HMCS samples present similar surface oxygen modification, structure property, and morphology but different surface pore sizes, making it an ideal research platform for studying the influence of mesoscale mass transport. It is revealed that in low current density conditions large surface area is preferred but the mass transport governs the performance in high current density region. On account of the favorable porous structure, HMCS-8 nm delivers the most excellent practical performance (166 mW cm-2) and performs well in the bifunctional Zn-air battery for the wastewater purification (70% RhB degraded after 2 min and 99% after 32 min).

Then we deepen our research to develop electrocatalysts for 2e– ORR in acidic media. We design a series of Au@Pd core@shell structures to investigate the influence of Pd 4*d* orbital overlapping degree on 2e– ORR performance. Density Functional Theory (DFT) calculations indicate that enhanced H2O2 selectivity and activity is achieved at Pdn clusters with $n \leq 3$, and Pd clusters larger than Pd3 should be active for 4e– ORR. However, experimental results show that Au@Pd NWs with Pd4 as the primary structure exhibit the optimal H2O2 performance in acidic electrolyte with high mass activity (5.71 A mg-1 at 0.4 V) and H2O2 selectivity (nearly 95%). Thus we report that Pd4, instead of Pd3, is the upper threshold of Pd cluster size for an ideal 2e– ORR. It results from the oxygen coverage on catalyst surface under potentials, and such oxygen coverage phenomenon causes electron redistribution and weakened *OOH binding strength on active sites, leading to enhanced activity of Pd4 with only 0.06 V overpotential in acidic media.

Lastly, the influence of local coordination environment (LCE) manipulation was studied. We underscore the influence of LCE on directing the 2e– ORR, utilizing Pd cluster as a model catalyst.

DFT calculations illustrate the role of first- and second-coordinated sulfur and oxygen in modulating the binding strength of oxygen intermediates. As guided by the DFT screening, the as-prepared Pdx/HMCS presents exceptional catalytic performance with a high mass activity of 4.06 A mg−1 at 0.45 V and selectivity above 94% over a wide potential range in 0.1 M HClO4 electrolyte. Based on the elaborated *in-situ* characterizations, we confirm that oxygen migrates from the carbon support (second coordination sphere) to the Pd cluster (first coordination sphere) to achieve oxygen coverage on the catalyst surface under potentials. As evidenced by DFT calculations, such an oxygen migration phenomenon as well as optimized first and second coordination environment give rise to outstanding performance.

703g Electrostatic Funnel, Ionic Concentration Polarization, and Diffusion Limited Ionic Transport in Charged Membranes

Joan Montes de Oca, Juan J. de Pablo

In this presentation, we will explore the characteristics of a phenomenon we have termed "Electrostatic Funnel" (EF) and its interplay with ion concentration polarization and diffusion-limited ionic transport, both well-established concepts in the field.

The EF is a non-equilibrium phenomenon that occurs as a manifestation of charge conservation at the boundary where charged membranes interface with their adjacent solutions. This scenario typically presents a disparity in the concentration of charge carriers on either side of the membrane, leading to a conductivity mismatch at this junction. Contrary to expectations, this mismatch cannot be corrected by merely accelerating the ions in the solution to maintain current continuity, as that would deplete the charge carriers neutralizing the membrane's interior. Instead, an electrostatic potential is droped at the membrane's entrance, creating an "Electrostatic Funnel." This mechanism effectively adjusts the ion concentration from the diluted external reservoir to the levels necessary for membrane neutralization. Remarkably, it achieves current continuity not by increasing the velocity of charge carriers but by augmenting their concentration. The influence of the EF thus hinges on various factors, including the electrolyte concentration in the reservoir, the membrane's charge density, and ratio of the ionic current sustained by the membrane and the reservoir.

Our findings also reveal that while the EF enhances the "effective" cross-sectional area of the membrane pores, it simultaneously creates a depletion zone of ions within. This latter effect, known as ionic concentration polarization, is recognized for its negative impact on the maximum current a membrane can support, constrained by diffusion-limited transport mechanisms.

703h Enhanced Fatty Alcohol Electrocatalytic Oxidation over Inorganic-Polymer Hybrid Interface

Ruiqi Du, Rui Jia, Kaizheng Zhang, Yi Cheng

Electrocatalytic alcohol oxidation presents a promising method for producing carboxylic acids, utilizing electrons as an environmentally friendly oxidizing agent. Fatty acids are an important class of fine chemicals widely utilized in cosmetics, plasticizers, resins, and other fields. Achieving selective oxidation of fatty alcohols stands as a pivotal route for fatty acid synthesis. Although electrocatalytic oxidation has shown remarkable efficacy in short-chain alcohols like methanol and ethanol, it encounters obstacles when extended to the oxidation of fatty alcohols with longer carbon chains. The extremely low solubility of fatty alcohols in aqueous electrolytes leads to mass transfer limitations, severely impeding the efficiency of electrocatalytic oxidation.

This work investigated the modulation of electrode-electrolyte interface to enhance fatty alcohol electrocatalytic oxidation, utilizing n-octanol as a model substrate. We employed a one-step electrodeposition method to fabricate polytetrafluoroethylene (PTFE)-decorated $Ni(OH)_2$ electrocatalyst. The inherent hydrophobicity of PTFE nanoparticles endowed this inorganic-polymer hybrid interface with higher affinity for n-octanol, facilitating its enrichment at the interface. Compared to pure $Ni(OH)_2$ electrocatalyst, PTFE-decorated $Ni(OH)_2$ catalyst exhibited significantly higher current density for n-octanol oxidation. The faradaic efficiency of n-octanoic acid product increased from 80% to 95% at 1.5 V vs. RHE, indicating that the hybrid interface not only promoted n-octanol mass transfer but also suppressed competing oxygen evolution reactions. This work sheds light on mass transfer enhancement through interface engineering strategies, with potential applicability to other electrocatalytic reactions involving water-insoluble substrates.

567a Smart Integration of Thermal Energy Storage for Grid Demand Response.

Mohamed Bahr, Jake Immonen, Blake Billings, Kody Powell

The manufacturing sector is actively exploring innovative approaches for mitigating peak demand penalties. With thermal energy storage (TES) experiencing significant advancements and diversifying in application, it has captured the attention of numerous industrial operations. Such technologies offer promising avenues for moderating peak electricity demands and, consequently, diminishing demand charges. This research endeavors to showcase the effectiveness of TES systems in reducing demand charges while emphasizing recent progress in integrating TES with intelligent control mechanisms. We provide a comparative analysis of energy use and peak demand in an industrial setting, both with and without TES systems, under a predefined charging and discharging regimen. Furthermore, the study delves into the benefits of deploying a smart control system capable of dynamically scheduling TES operations based on real-time energy consumption and preset objectives. Findings reveal significant financial benefits stemming from TES utilization across two distinct scenarios, manifesting in lowered energy costs for the facility under investigation. Specifically, integrating a smart controller with TES is projected to yield annual savings of approximately 3.3% in energy charges, excluding potential incentives. By contrast, savings accruing from a regimented charging schedule are estimated at 2.7% annually.

567b Phase Change Material Heat Exchanger to Decarbonize Water Heating in Low-Income Housing: Justification and Heat Exchanger Design

Erin Blackley

In the global push towards decarbonization, the buildings sector is particularly well-positioned to contribute by offering many opportunities for deep electrification. Of existing thermal systems in residential buildings, water heating (traditionally powered by fossil fuels) currently accounts for about 20% of household energy consumption in the US. Electric water heaters have grown in popularity in recent years, comprising nearly 50% of existing US water heating systems; however, a majority of these units are electric resistance water heaters (ERWHs), which suffer from low efficiency, high energy consumption, and sometimes require an electrical panel upgrade, making electrification difficult in low-and-moderate income housing. Heat pump water heaters (HPWHs), another electrified alternative, use 60-70% less energy than ERWHs and can therefore often be installed without requiring panel upgrades. However, deployment of HPWHs has been hindered in space-constrained, low-income housing such as multifamily buildings or mobile/manufactured homes, as HPWHs tend to be significantly larger than traditional water heaters and cannot fit into the small utility closets often present in these settings. Currently, 32% of all housing in the US consists of multifamily or mobile homes, and of these units, over 60% have water heaters less than 50 gallons. Therefore, there is a need to develop smaller 30 – 40-gallon HPWHs with either equivalent or improved performance compared to larger electric or gas water heaters.

This presentation will first describe the overall need and efforts to incorporate thermal energy storage (TES) into small-volume HPWHs to meet space requirements in low-income housing while maintaining hot water requirements. A central part of the proposed solution is a novel phase change material (PCM)-integrated heat exchanger that increases the TES capacity of a HPWH, effectively allowing improved performance with smaller water tanks. The primary focus detailed here lies in the design development of a heat exchanger with thermal storage capabilities with respect to HPWH operation and associated water tank models. Using generative design and bio-inspired lattice geometries like triply periodic minimal surfaces such as gyroids, our efforts show great promise for enhancing heat transfer of the PCM heat exchanger, enabling rapid PCM charging and discharging that maximizes performance of small-volume HPWHs for low-income housing. When comparing a standard ERWH to a smaller HPWH that contains a PCM-integrated heat exchanger, preliminary results show that the latter system can overcome the heat storage capacity shortcomings of the ERWH tank with a heat exchanger comprising less than 10% of the HPWH tank volume. Our current modeling efforts show that a 40-gallon equivalent HPWH with 50% volume occupied by the PCM heat exchanger can at least double the heat storage and hot water delivery capabilities of an ERWH while maintaining low heat exchanger production costs.

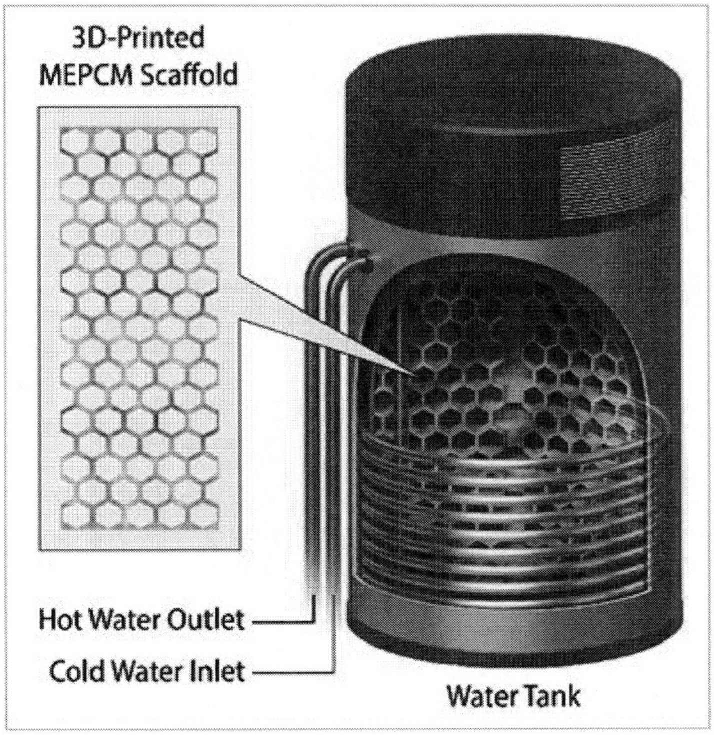
3D-Printed
MEPCM Scaffold
Hot Water Outlet
Cold Water Inlet
Water Tank

567c Phase Change Material Heat Exchanger to Decarbonize Water Heating in Low-Income Housing: Material Engineering Considerations

Chase Robinson

In the global push towards decarbonization, the buildings sector is particularly well-positioned to contribute by offering many opportunities for deep electrification. Of existing thermal systems in residential buildings, water heating (traditionally powered by fossil fuels) currently accounts for about 20% of household energy consumption in the US. Electric water heaters have grown in popularity in recent years, comprising nearly 50% of existing US systems; however, a majority of these water heaters are electric resistance water heaters (ERWHs), which suffer from low efficiency, high energy consumption, and sometimes require an electrical panel upgrade, making electrification difficult in low-and-moderate income housing. Heat pump water heaters (HPWHs), another electrified alternative, use 60-70% less energy than ERWHs and can therefore often be installed without requiring panel upgrades. However, deployment of HPWHs has been hindered in space-constrained low-income housing such as multifamily buildings or mobile/manufactured homes, as HPWHs tend to be significantly larger than traditional water heaters and cannot fit into the small utility closets often present in these settings. Currently, 32% of all housing in the US consists of multifamily or mobile homes, and of these units, over 60% have water heaters less than 50 gallons. Therefore, there is a need to develop smaller 30 – 40-gallon HPWHs with either equivalent or improved performance compared to larger electric or gas water heaters.

This project details the fabrication of a phase change material (PCM) heat exchanger for use in small-volume HPWHs to increase thermal storage capacity and overall performance. The design of the novel heat exchanger relies on additive manufacturing (3D printing) to produce complex geometries not possible via traditional manufacturing methods. This presentation describes efforts toward material processing and performance characterization for the 3D-printed PCM heat exchanger as well as preliminary results showing the applicability of additive manufacturing for introducing thermal energy storage into a HPWH. Printing techniques such as fused deposition modeling (FDM) pair well with micro-encapsulated PCMs (MEPCMs) and take advantage of low-cost, accessible methods and materials such as polymers. Based on standards for approved polymers in potable water systems, this work uses high-density polyethylene (HDPE); however, due to the novelty of PCM heat exchangers within domestic water heaters, there is a lack of regulation guidance for PCM selection. Therefore, our PCM selection relied on available MEPCMs with low toxicity, high energy storage density, and an appropriate transition temperature for water heating applications. Based on preliminary HPWH modeling results, our work utilizes an MEPCM with a transition temperature of 55°C. The primary goal of this work is to develop an MEPCM-HDPE composite with the highest possible MEPCM content that allows for reliable 3D printing and does not compromise the heat exchanger's structural integrity or durability. A higher MEPCM content corresponds to a heat exchanger with a higher thermal energy storage density, thus providing greater benefit for small-volume HPWHs. Preliminary efforts show the production of composite filaments of at least 60% MEPCM by weight with uniform consistency and properties, as well as small-scale heat exchanger prototypes produced via FDM. This work stands as a promising a path forward towards achieving superior HPWH performance in space-constrained, low-income housing.

567d PCM-Based Thermal Energy Storage to Enable Cold Climate Operation of R290 Residential Heat Pump

Sarath Kannan, Kenneth Hultquist, Carlos Ortiz, Kaushik Biswas

The technical and economic feasibility of integrating an R290 heat pump (HP) with a low-cost-multi-temp salt hydrate-based PCM thermal energy storage (TES) system to reduce equipment size, and initial and operating costs associated with space conditioning (Space Heating + Space Cooling) and water heating for cold climates is explored through experimentation and modelling.

In addition to having good thermodynamic properties, R290 refrigerants are also environmentally friendly (GWP < 3) compared to traditional R-410A (GWP > 2,000). However, due to flammability concerns their charge size is regulated to 150 grams for residential applications, limiting their capacity. This can be overcome by leveraging thermal energy storage solutions. In the experimental investigation, a residential scale R290 heat pump (16 kW) is hydronically coupled with a multi-temp salt-hydrate-based thermal energy storage system (T_m = 30, 40, 60 °C). The multiple temperatures allow for optimized operation to meet space and water heating demands in addition to utilizing the economic benefits of forward capacity and demand flex market. Subsequently, the performance of the heat pump and the HP-TES system is evaluated for different operating conditions. The former set of experiments help characterize the performance and efficiency of the heat pump for different ambient (T_{amb} = -15 to 40 °C), leaving water temperatures (T_{LWT} = 7 to 65 °C) and modulation ratios and the latter is used to evaluate the same for the HP-TES system. Using these empirical data, simulation models are built to assess critical cost and efficiency matrices under different operating parameters and weather scenarios to demonstrate and facilitate its adoption for cold climate zones.

567e Performance Study of High Temperature Microencapsulated Phase Change Material for Thermal Energy Storage through Mathematical Modeling

Sk. Moumita Manjur, Sidney Lin

Heat produced during industrial processes can be captured and reused to minimize the amount of heat loss. Latent heat storage using phase change material (PCM) is a promising technique to capture and recycle industrial waste heat. PCM can be used to capture different grades of waste heat. Use of PCM to capture high temperature (> 400 °C) waste heat is under development. A mathematical simulation can help understand the heat transfer behavior and efficiency to optimize the heat storage system design and operation. Microencapsulated phase change materials (MEPCMs) consisting of phase change materials packed inside a resilient shell (e.g., polymer, alumina, sodium silicate, or steel) have advantages of isolation from the external environment and preventing materials loss by liquid leakage. Additionally, MECPCM can easily be transported from the charging site (industrial facility) to the service site. This work studies the heat transfer and phase change behaviors of high temperature MEPCM during charging (heating) and discharging (cooling) processes.

A finite element analysis model of heat transfer and solid-liquid phase change in a phase change material has been developed using COMSOL Multiphysics®. In this model, temperatures at different time, location, and PCM phase (solid or liquid) inside the MEPCM of a shell have been calculated. Different materials have been used as PCMs to study their individual phase change behavior and heat storage capacity. Alumina was used as the outer shell material of the MEPCM. Temperatures of MEPCM at solid and liquid phases were calculated using Fourier's law of heat conduction integrated with heat convection between the MEPCM and surroundings. Temperature during solid/liquid phase transition was calculated numerically using the general concept of first order phase transition by absorbing/releasing a fixed amount of heat at a constant temperature. Temperature and composition dependent thermophysical and thermodynamic properties of the PCMs and alumina from literature were used in this model. Three different heating/cooling conditions were studied in this work: 1) the outermost surface of the MEPCM is maintained at a constant temperature; 2) a constant heat flux is applied on the MEPCM shell surface and 3) a convection heat transfer condition is used for heat exchange between environment and MEPCMs. Different environments have been used for convection heat transfer conditions, including air, water, and oil. The calculated results were analyzed to study the phase change phenomena of different MEPCMs in different conditions and environments, which reflect comparative understanding about the overall performance of MEPCM heat storage capacity and heat transfer rate during the charging and discharging processes.

567f Design of a Pumped Thermal Energy Storage (PTES) System Capable of Delivering High-Grade Heat

Alp Albay, Mehmet Mercangöz

Pumped thermal energy storage (PTES), which stores electrical power as heat, is a promising approach for long-duration energy storage. PTES systems can enhance the reliability, stability, and effectiveness of renewables-rich power systems. While charging, a PTES system runs as a heat pump with surplus electricity, which: (i) supplies heat to a hot store, creating a high-temperature reservoir, and (ii) producing a cooling effect in a cold store, forming a low-temperature reservoir. When electricity is needed, the PTES system can be reversed and function as a heat engine. This technology can address not only the inherent intermittency of renewable sources, which most long-term energy storage technologies can address, but also address the need for decarbonizing heat. Heat demand comprised 48.7% of the global energy demand in 2020 but renewables accounted for just 12% of this heat demand [1][2]. This highlights the need for research into expanding decarbonized heat supply. In this work, we propose an augmented PTES system, capable of delivering both electricity and heat while maintaining consistent operation in both modes. This is accomplished through careful management of the state-of-charge of the storage media and addition of balancing heat exchangers. The basis of the PTES system lies in a supercritical CO_2 Brayton cycle which has been investigated thoroughly in recent studies [3].

The system is shown in Figure 1a. The overall process was built on previous work by the authors is explained in greater detail in [4] and [5]. The methodology of the process design involves implementing state-of-charge management into each cycle's design. The goal of this is to maintain equal SoC at both the hot and cold storage tanks. This is done through an algorithm outlined in Figure 2. When the system is designed according to this flowchart, the rate of change in both tanks are always equal. However, this only holds for electrical charge and discharge. When the system is used to supply heat from the hot reservoir, this balance is broken. Thus, there is a need for maintaining this balance externally. This is done in through a balancing heat exchanger that uses excess hot storage media to heat up unused cold storage. The augmented process is shown in Figure 1b, for a side-by-side comparison with the base case. This process can then be used to provide heat to a coupled industrial process, provide district heating or the heat can be sold. Heat integration increases the overall efficiency of the storage plant by 35% while also providing renewable-based heat supply.

References

[1] REN21, "Renewables 2023 Global Status Report," REN21, Paris, 2023.

[2] IEA, "Renewable heat – Renewables 2022 – Analysis," IEA, 2022. https://www.iea.org/reports/renewables-2022/renewable-heat.

[3] Marchionni, G. Bianchi, S. A. Tassou, Review of supercritical carbon dioxide (sco2) technologies for high-grade waste heat to power conversion, SN Applied Sciences 2 (03 2020). doi:10.1007/s42452-020-2116-6.

[4] Zhu, A. Albay, M. Mercangoz, "Pumped Thermal Electricity Storage: Modelling of a Reversible Recuperative sCO2 Pumped Brayton Cycle for Variable Power Operation", presented at the AIChE Annual Meeting, Orlando. FL, November 2023

[5] A. Albay, Z. Zhu, and M. Mercangöz, "State-Of-Charge (SOC) Management of PTES Coupled Industrial Cogeneration Systems," in *Turbo Expo 2024*, ASME, Jun. 2024.

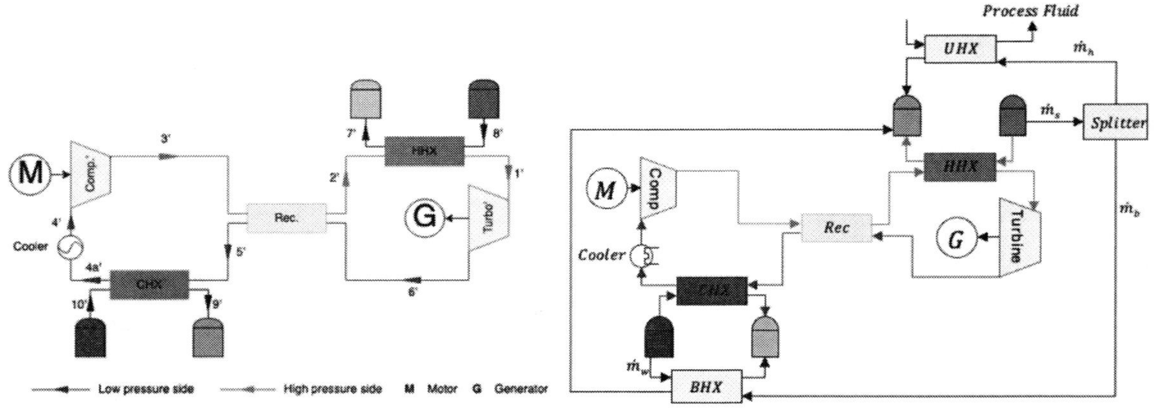

Figure 1: Discharging cycle of the PTES system. Left: Base design. Right: Design for heat integration.

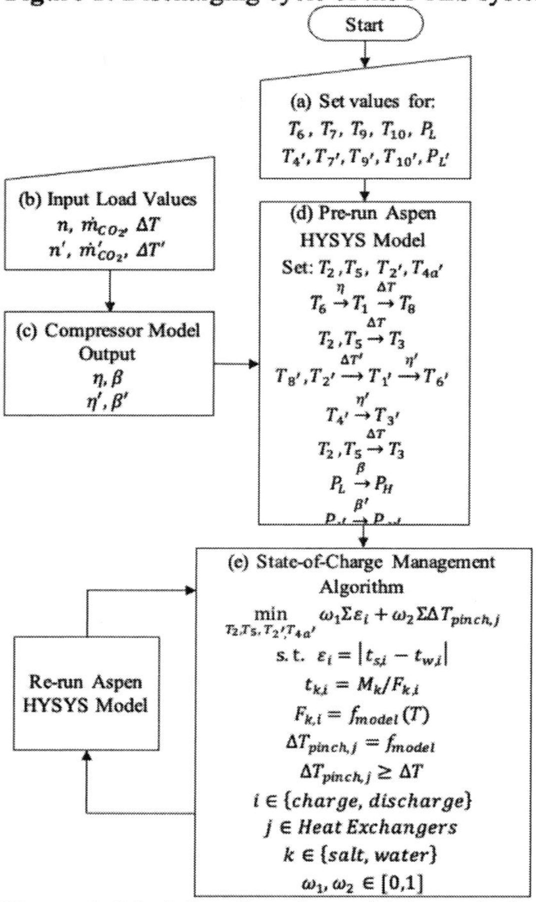

Figure 2: Modeling flowchart with in-cycle SoC management algorithm.

567g Design and Demonstration of a High Energy Density Hydrogel-Based Thermal Energy Storage Device

Bachir El Fil, Xiangyu Li, Carlos Diaz, Liliosa Cole, Joseph Mooney, David Keisar, Hyeongyun Cha, Gustav Graeber

Thermal energy storage (TES) technologies are integral components in the shift towards more sustainable and efficient energy systems, playing a pivotal role in managing the supply and demand of energy, particularly from renewable sources. Among the various TES methods, such as sensible heat, latent heat, and thermochemical energy storage, the latter stands out due to its superior energy storage density and the ability to store heat for extended periods without noticeable losses. Thermochemical energy storage (TCES) devices harness chemical reactions to absorb or release thermal energy, offering a groundbreaking solution to the intermittent challenges of renewable energy sources. Conventional materials used in TCES systems, including metal hydrides, zeolites, and salt hydrates, are chosen for their favorable reaction enthalpies, which significantly influence the system's energy density. These materials can store and release large amounts of heat through reversible chemical reactions, making TCES systems highly efficient and compact compared to other TES technologies. The energy density of TCES systems is higher than that of sensible and latent heat storage methods, allowing for more energy to be stored in a smaller volume i.e., significantly higher energy densities, thereby enhancing the feasibility and efficiency of renewable energy systems. This unique advantage positions thermochemical energy storage as a key technology in the development of future energy storage solutions, offering a sustainable path towards meeting global energy demands. Despite having many advantages, TCES systems face several challenges and shortcomings that result in impeding their adoption of residential and commercial applications. These shortcomings not only encompass material and system-level challenges but also directly influence the energy input required for recharging and the overall round-trip efficiency of the systems. Addressing these issues is crucial for advancing TCES technologies and ensuring their viability in the broader energy market.

To overcome these shortcomings, we have developed and optimized a novel thermal storage material with tunable performance *i.e.*, uptake and kinetics. The proposed TES device utilizes a specially engineered hydrogel material that exhibits a tunable phase change temperature, desorption temperature, water uptake, and significant latent heat storage capacity. Hydrogels, with their unique network structure capable of holding large amounts of water, have been modified to undergo a solid-liquid phase transition at a predetermined temperature, suitable for TES applications. This phase change is associated with the absorption or release of substantial thermal energy, enabling efficient storage and release of heat. The hydrogel's chemical composition and physical properties have been optimized to maximize energy storage density and thermal conductivity while ensuring stability and durability over repeated thermal cycling (> 200 cycles). We overcome the limitations of current solutions using a novel hydrogel/salt composite as the adsorbent, reaching higher energy density with lower discharging temperature ranging from 50-80°C, while offering the benefits of low cost and scalable fabrication. The hydrogel is superior to most other storage media except for metal organic frameworks (MOFs) that can typically be very expensive and require high regeneration/desorption. After incorporating hygroscopic salts into a hydrogel matrix, the matrix can prevent the hydrated salts from deliquescence, which maintains the outstanding moisture absorbing capability of salts, over >1.8 gram of water per gram of polymer at 30%RH, and stable cyclability with no hysteresis. The adoption of hydrogel for TES combines the benefits of high energy density and low discharging temperature, both of which are critical to reducing the cooling cost in residential and commercial buildings. Using a PAM-LiCl (4 g LiCl per gram of polymer) resulted in a system with record high energy density > 200 kWh/L. The findings and innovations detailed in this work have the potential to revolutionize the energy sector, fostering the transition to a more sustainable and resilient energy future.

444a Engineering 2D Materials for Zn-Ion Batteries and Na-Ion Batteries

Aiping Yu

To tackle the environmental challenges, renewable energy storage including batteries and supercapacitors has garnered great attention over the past decade. My group has been focused on the production and modification of graphene, MXene, and other nanomaterials for Na-ion batteries, and Zn-ion batteries. Our strategy for nanomaterials control includes but is not limited to novel production methods, morphology control, chemical functionalization/doping, and composite structure. The produced new structured materials have been proven to dramatically enhance the performance of the relevant batteries.

444b Micron Size $NaCrO_2$ Particles Enable High Loading Dry-Process Electrode for Sodium Ion Batteries

Junlin Wu, Zheng Chen

Traditional sodium-ion battery (SIB) electrode fabrication typically relies on a wet slurry process involving N-methyl-2-pyrrolidone (NMP), a solvent known for its high cost, environmental harm, and toxicity. As an alternative, dry-process fabrication using polytetrafluoroethylene (PTFE) is emerging as a preferred method due to its cost efficiency, eco-friendliness, and potential to achieve higher energy densities. While numerous studies have demonstrated the high loading compatibility of the dry process method compared to the traditional slurry method, there remains a gap in understanding the critical material parameters necessary for producing high mechanical property films and enhancing their electrochemical performance. In this study, two different sizes of $NaCrO_2$ particles noted as S-NCO (300-600 nm) and L-NCO (5-20 μm) were synthesized to investigate the impact of particle size on dry film fabrication. Our findings reveal that electrodes made with larger L-NCO particles exhibit significantly higher high tensile strength of around 24 kPa over S-NCO film (less than 10 kPa) at the same PTFE binder ratio, as well as enhanced electronic conductivity and electrochemical performance when the same carbon ratio is added. These improvements are primarily attributed to the low porosity of L-NCO film induced by higher tapped density and reduced surface area of L-NCO powder. Notably, a full cell incorporated with a high-loading L-NCO film electrode (45 mg/cm^2 or 5.4 mAh/cm^2) demonstrates exceptional cycling performance (80% capacity retention over 300 cycles) compared with S-NCO film electrodes, highlighting the potential advantages of using larger particle sizes in dry film electrode fabrication. These results provide insights regarding the design and fabrication of PTFE-based cathode dry-process film electrode and promote the high-loading and high-performance electrodes for future energy storage applications.

ACS Energy LETTERS

http://pubs.acs.org/journal/aelccp

Colossal Capacity Loss during Calendar Aging of Zn Battery Chemistries

Bo Liu, Xintong Yuan, and Yuzhang Li*

Cite This: *ACS Energy Lett.* 2023, 8, 3820−3828 Read Online

ABSTRACT: Whereas previous research efforts in Zn battery chemistries have primarily focused on extending their cycle life, calendar aging has largely been neglected and is poorly understood. Here, we discover that Zn metal anodes lose 12−37% of their capacity after only 24 h of calendar aging, which is more than an order of magnitude greater than any other battery chemistry previously studied. We find that this large capacity loss occurs regardless of electrolyte chemistry, substrate composition, and calendar aging time, suggesting that the losses cannot be accounted for by corrosion alone. By leveraging titration gas chromatography, we distinguish corrosion losses from losses due to electrically disconnected Zn ("dead" Zn) and quantify dead Zn as the main contributor to capacity loss during Zn battery aging. More broadly, this study shows that large instabilities in seemingly nonreactive metals (e.g., Zn) can occur, highlighting the importance of understanding their underlying degradation mechanisms.

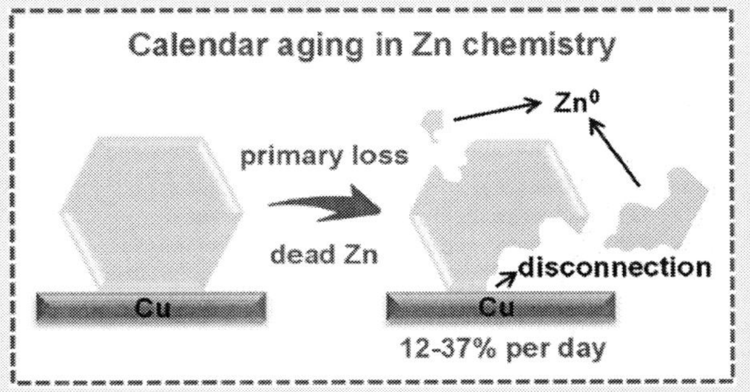

Many drawbacks of today's dominant Li-ion battery technologies (e.g., high cost, flammability, critical minerals) can be mitigated by Zn-based battery chemistries.[1] Correspondingly, low-cost Zn batteries are a promising alternative for practical applications at the grid-scale level, which require long cycling stability and calendar life (i.e., long rest periods at open-circuit potential), among others.[2,3] While improvements in cycling stability have been demonstrated through innovations in nanostructuring[1,4−6] and electrolyte chemistry,[7−10] capacity retention during calendar aging is equally important but much less studied or understood. Such rest periods are ubiquitous in typical battery operation (e.g., parked cars for electric vehicles and intermittency during grid storage) yet have not been thoroughly quantified for Zn battery chemistry. Indeed, corrosion processes during calendar aging can be triggered by the low electrode potential of Zn and the narrow electrochemical stability of aqueous electrolytes.[1,11−14] A critical need remains to quantify these corrosion effects on capacity retention and elucidate their nanoscopic origins and mechanisms in order to mitigate their impact.

The calendar aging mechanisms in Li-based battery anodes (e.g., graphite, Si, Li metal) are well known to proceed through electrolyte reduction at the anode surface[15−19] to continuously form the solid electrolyte interphase (SEI) (Figure 1a). Although this surface corrosion film provides some degree of passivation, electrolyte infiltration through the SEI may cause the observed SEI thickening during calendar aging, which is driven by the electrode potential and the voltage stability window of the liquid electrolyte.[20] For graphite and silicon, the capacity loss can be as little as 1% fade per month, especially when aged in the partially delithiated (i.e., discharged) state at higher electrode potentials.[15,16,21−25] For metallic Li, SEI thickening is accelerated because the electrode potential of the Li metal is always at 0 V vs Li/Li^+, which results in a significant 2% capacity fade or more during the first 24 h.[17,18,20] These previous studies on Li-based anodes rely on coulometry and directly attribute any measured capacity losses to corrosion alone. However, this methodology fails to capture the contribution of any inactive material that could be formed during calendar aging, which can be particularly significant for Zn-based batteries. Recently, Pu et al. explored the Zn-anode losses that occur during calendar aging. Indeed, the capacity loss was shown to form in only one specific system: Zn/Ti half-cells with mildly acidic 2 M zinc sulfate ($ZnSO_4$) as the electrolyte.[26] Furthermore, the few calendar aging studies on Zn-based anodes involving full-cells primarily examined

Received: June 27, 2023
Accepted: August 16, 2023
Published: August 21, 2023

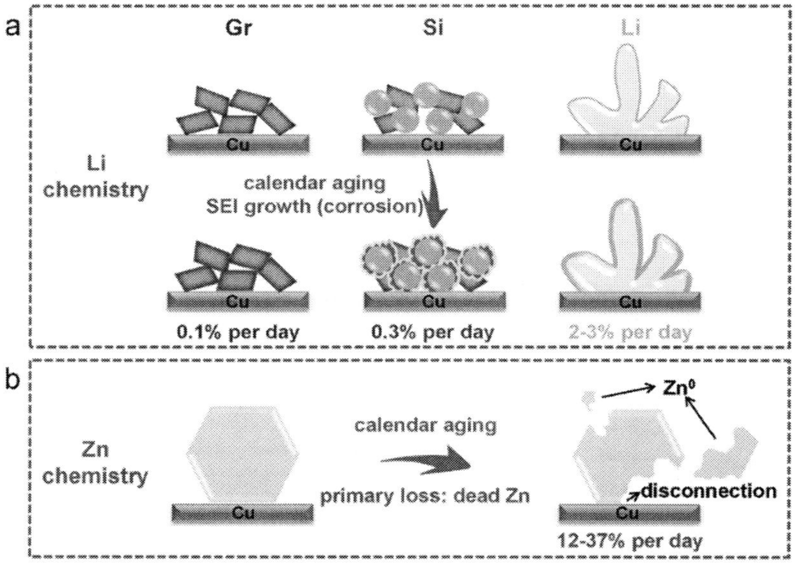

Figure 1. Different mechanisms of calendar aging between Li chemistry and Zn chemistry. (a) Continuous SEI formation (corrosion) through electrolyte reduction at the anode surface during calendar aging in Li-based battery anodes (e.g., Gr, Si, and Li metal). (b) Dead Zn accounts for the large majority of capacity losses during calendar aging in Zn chemistry.

capacity losses with Zn-excess full-cells, which failed to isolate the effects of anode and cathode on aging.[9,27−29] Therefore, the detailed aging mechanism of the metallic Zn anode remains elusive, representing a key gap in our systematic understanding of Zn battery chemistry. The lack of a conformal SEI on Zn metal and the narrower voltage stability window of the aqueous electrolyte[12,30,31] may result in uniquely distinct calendar aging mechanisms for these battery chemistries, making them important to elucidate and quantify.

Here, we discover a colossal capacity loss of 12−37% during 24 h calendar aging (regardless of the electrolyte chemistry or current collector substrate) in Zn metal anodes, a value that is an order of magnitude greater than any other battery chemistry previously measured (Figure 1b). The cumulative effects of corrosion and dead Zn formation are shown to be detrimental to full-cell cycling stability in an anode-free configuration. Using electron microscopy, we reveal that the interface between Zn and the current collector substrate deteriorates through corrosion, suggesting possible mechanisms for the formation of significant dead Zn. This shows that Zn battery calendar aging mechanisms are distinct from Li battery chemistries; attributing the colossal coulometric capacity losses to corrosion alone would fail to consider the contribution of dead Zn to calendar aging. Therefore, we use titration gas chromatography (TGC) to quantitatively differentiate corrosion products (i.e., Zn^{2+}) from dead Zn formation (i.e., Zn^0), the resulting inactive Zn of which accounts for the large majority (e.g., more than 75%) of capacity losses during calendar aging. Together, our multiscale characterization study provides a holistic picture of the important calendar aging processes in Zn battery chemistries and identifies potential insights to mitigate its impact.

Parameters Impacting Calendar Aging of Metallic Zn. During calendar aging, side reactions within the resting battery continue to occur at the open-circuit potential. Capacity losses during calendar aging can be quantified as the difference between a battery's Coulombic efficiency (CE) when cycled without a rest period and a battery's CE when cycled with a specified rest period (e.g., 24 h) at open-circuit voltage.[17,18,32] This is termed ΔCE or CE loss. Previous work using this protocol showed capacity losses during calendar aging in the range of 0.1% CE loss per day (i.e., graphite[16,22,24,27]) to 2−3% CE loss per day (i.e., Li metal[21]) and attributed these losses to corrosion (Figure 1a). Here, we age Cu|Zn half-cells after Zn electrodeposition in the second cycle at open-circuit voltage under a variety of conditions (i.e., electrolyte chemistry, substrate architecture, deposited capacity, temperature, aging time) and calculate the capacity loss during calendar aging as described above following the formula CE loss = $CE_{without\ aging} - CE_{with\ aging}$.

Electrolyte Chemistry. We quantify CE loss during 24 h calendar aging in a diverse range of electrolyte chemistries that span different salt configurations (bulky anion vs small anion), salt concentrations (high vs low concentration), and salt additives (e.g., $MnSO_4$). These electrolyte chemistries all exhibit varying CE values and stabilities during typical galvanostatic cycling experiments previously described in the literature[2,17,33] (Supplementary Fig. 1). When examining such electrolytes for calendar aging, we measure significant CE losses between 12 and 37% (Figure 2a and Supplementary Fig. 2), which is an order of magnitude greater than previously reported Li battery chemistries[3] and agrees with a previous study on Ti|Zn half-cells in 2 M $ZnSO_4$ electrolyte.[26] It is interesting to highlight here that, while electrolyte chemistry can dramatically impact the CE during galvanostatic cycling, with values spanning 50−99%, large calendar aging losses occur regardless of electrolyte chemistry and are at least an order of magnitude greater than those of previously tested metallic Li anodes.

To further study the effect of substrate architecture, deposition capacity, temperature, and aging time on CE losses during calendar aging, we quantify ΔCE in a model electrolyte

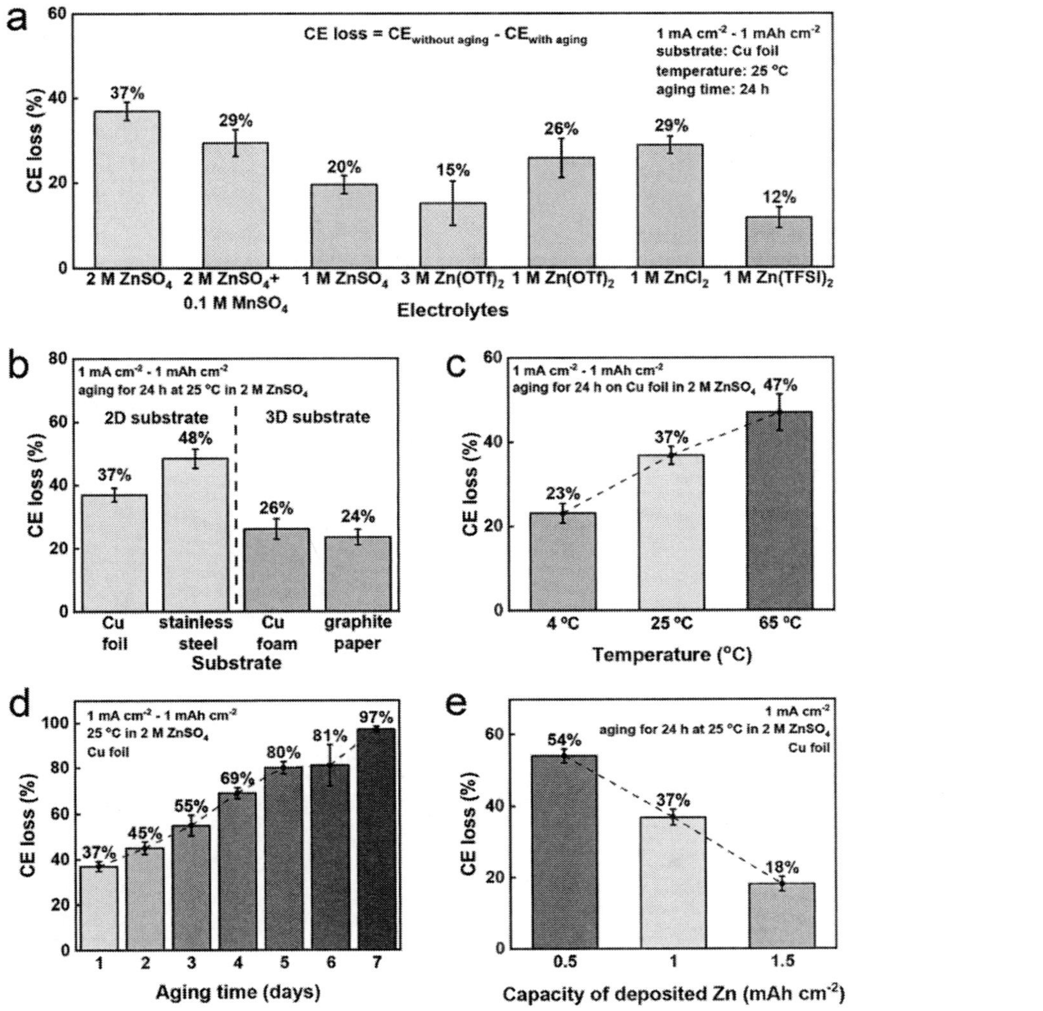

Figure 2. Calendar aging effects on the CE of Zn metal anodes. (a) CE losses of Zn metal in various electrolytes, including 2 M $ZnSO_4$, 2 M $ZnSO_4$ with 0.1 M $MnSO_4$ as additives, 1 M $ZnSO_4$, 3 M $Zn(OTf)_2$, 1 M $Zn(OTf)_2$, 1 M $ZnCl_2$, and 1 M $Zn(TFSI)_2$ (current density: 1 mA cm^{-2}, capacity: 1 mAh cm^{-2}, substrate: Cu foil, temperature: 25 °C, aging time: 24 h). Effects of (b) substrates (current density: 1 mA cm^{-2}, capacity: 1 mAh cm^{-2}, temperature: 25 °C, aging time: 24 h, electrolyte: 2 M $ZnSO_4$), (c) temperatures (current density: 1 mA cm^{-2}, capacity: 1 mAh cm^{-2}, substrate: Cu foil, aging time: 24 h, electrolyte: 2 M $ZnSO_4$), (d) calendar aging time (current density: 1 mA cm^{-2}, capacity: 1 mAh cm^{-2}, substrate: Cu foil, temperature: 25 °C, electrolyte: 2 M $ZnSO_4$), and (e) capacity of deposited Zn (current density: 1 mA cm^{-2}, substrate: Cu foil, temperature: 25 °C, aging time: 24 h, electrolyte: 2 M $ZnSO_4$) on CE loss. Each bar represents three data points or coin cells.

(2 M $ZnSO_4$) while independently varying each of the parameters above. For the **substrate architecture**, we compare the ΔCE in two-dimensional (2D) current collectors against that of three-dimensional (3D) structures. We observe a ~10% improvement in CE loss and a ~0.1 mAh decrease in capacity loss in 3D architectures (Figure 2b, e, Supplementary Fig, 3a,d,e, and Supplementary Fig. 4), as evidenced by the 36.9% ΔCE on 2D Cu foil and 26.1% ΔCE on 3D Cu foam under 1 mA cm^{-2} with a 1 mAh cm^{-2} capacity of deposited Zn (53.9% vs 43.2% for 0.5 mAh cm^{-2} and 18.1% vs 11.9% for 1.5 mAh cm^{-2}). Such an improvement might be attributed to the dramatically higher surface area of 3D substrates, leading to additional electrical pathways in 3D structures that can lessen inactive Zn formation[5,17,34] and reduce the anode impedance,[4,35,36] providing a potential guiding principle to minimize calendar aging losses. The impact of **temperature** on calendar aging appears to be large as well, with a CE loss of 23% when aged at 4 °C compared with a CE loss of 47% when aged at 65 °C (Figure 2c and Supplementary Fig. 3b). When varying **aging time**, we find that CE losses increase almost linearly with time, and nearly all capacity is lost after 7 days of calendar aging (Figure 2d and Supplementary Fig. 3c). Importantly, this continuous and rapid aging behavior, whether at low (1 mA cm^{-2}) or high current density (10 mA cm^{-2}),[26] is completely distinct from all other Li-based battery chemistries, as the capacity loss during aging in Li-ion anodes (i.e., graphite, Si, Li metal) levels off as the SEI layer becomes more passivating as it thickens during aging.[6] Finally, the CE loss

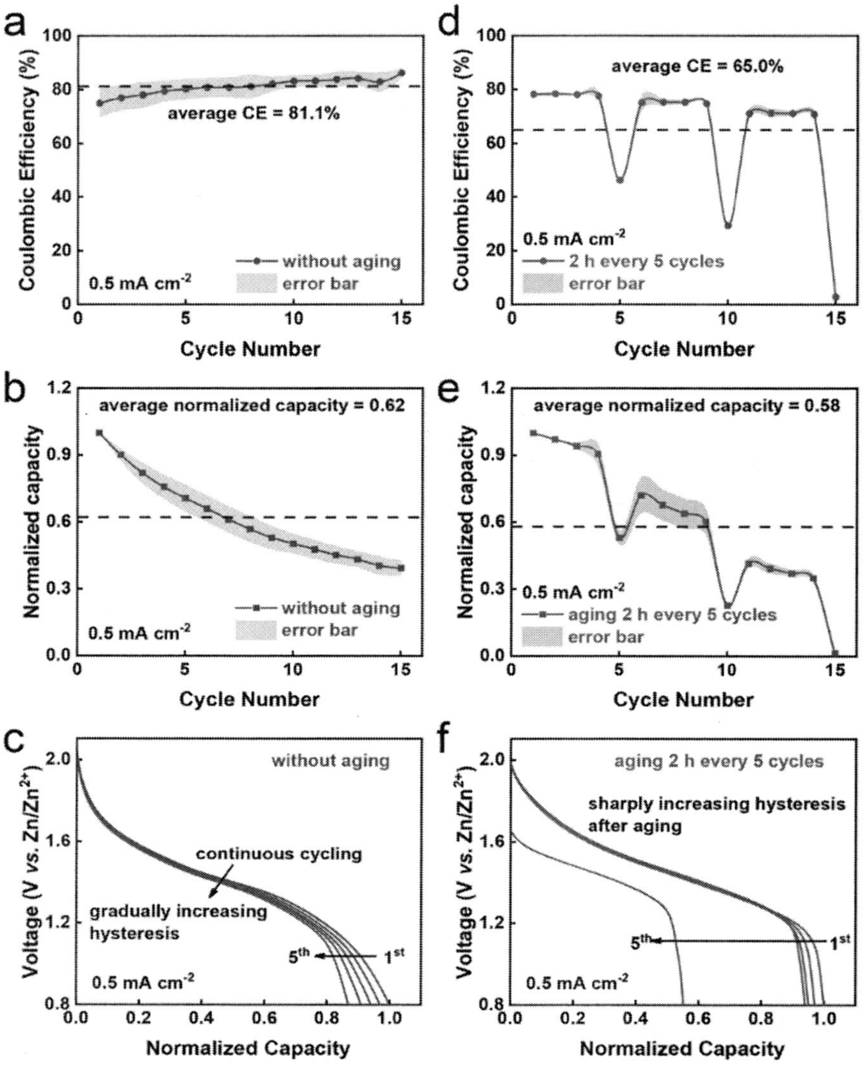

Figure 3. Calendar aging effects on the cycling performance of anode-free full-cells. (a) CE, (b) normalized discharge capacities, and (c) galvanostatic charge−discharge profiles of full-cells without aging. (d) CE, (e) normalized discharge capacities, and (f) galvanostatic charge−discharge profiles of full-cells with calendar aging every 5 cycles. Each point in (a), (b), (d), and (e) represents three data points or coin cells.

percentage appears to decrease with the increasing **deposition capacity** of metallic Zn (Figure 2e, Supplementary Fig. 3d, and Supplementary Fig. 4a,b). However, when converting the CE loss percentage to an absolute capacity loss in mAh, it appears that approximately the same amount of capacity is lost (~0.3 mAh for Cu foil and ~0.2 mAh for Cu foam) when changing Zn deposition capacity from 0.5 to 1.5 mAh cm^{-2} (Supplementary Fig. 3e and Supplementary Fig. 4a,b). This suggests that the absolute capacity loss during calendar aging is independent of the amount of Zn deposited.

Impact of Calendar Aging on Anode-Free Full-Cell Performance. In addition to half-cell experiments that isolate the detrimental effect of calendar aging on the Zn metal anode, we further quantify this impact for real battery performance. In particular, cycling experiments using anode-free full-cells are necessary, as these configurations maintain the same anode process of Zn electrodeposition onto the Cu current collector as in our half-cell experiments.[15,37,38] As a representative example of an anode-free full-cell battery,[39] we use a $ZnMn_2O_4$ (ZMO, Supplementary Fig. 5) cathode and a Cu foil as an anode substrate in 3 M $Zn(OTf)_2$. Capacity loss mechanisms during calendar aging in the 100% charged state will then be dominated by the electrodeposited Zn (Supplementary Fig. 6). During a standard galvanostatic cycling protocol without calendar aging, full-cells exhibit an average CE of 81.1%, with an average normalized capacity of 0.62 over 15 cycles (Figure 3a,b), which is typical of this anode-free configuration.[37] In contrast, anode-free cells aged once every 5 cycles for just 2 h in the charged state (i.e., electrodeposited Zn) demonstrate a marked decrease in the average CE (65.0%) and normalized capacity (0.58) over 15 cycles (Figure 3d,e). Strikingly, the normalized capacity of the 15th cycle falls to zero during our

https://doi.org/10.1021/acsenergylett.3c01282
ACS Energy Lett. 2023, 8, 3820−3828

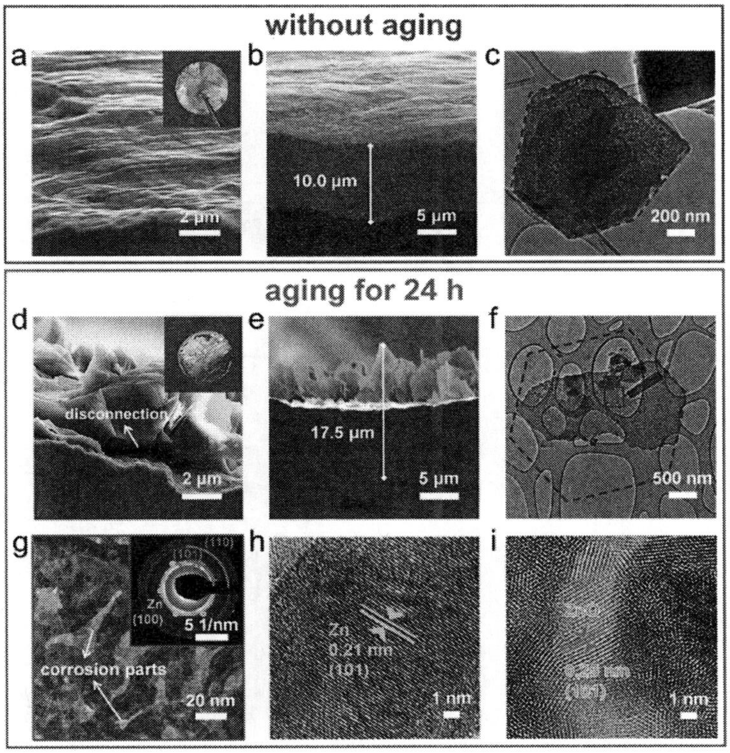

Figure 4. Nanoscopic structural evolution of inactive Zn species remaining on the Cu surface during calendar aging. SEM images of (a) the Cu surface and (b) the Cu thickness without aging. (c) TEM image of deposited Zn without aging. SEM images of (d) Cu surface and (e) Cu thickness after aging for 24 h. Insets in (a) and (d) are the corresponding OM images. (f, g) TEM and (h, i) HRTEM images of Zn residue left in the Cu current collector after aging for 24 h. Inset in (g) is the corresponding SAED image.

calendar aging protocol, highlighting the substantial impact of just three short aging periods (i.e., 2 h) on full-cell cycling performance. A higher frequency of 2 h aging every 3 cycles shows an even lower average CE of 52.5% and an average normalized capacity of 0.58 over just 6 cycles (Supplementary Fig. 7a,b). Furthermore, the voltage profiles of aged anode-free cells show a large hysteresis value of ~0.4 V after the fifth cycle (Figure 3f and Supplementary Fig. 7c), demonstrating that both voltage and capacity significantly decline during the rest period. Without calendar aging, the charge and discharge voltage profiles essentially overlap with no hysteresis (Figure 3c). These results underscore the importance of considering calendar aging when evaluating the performance of Zn battery chemistries, with merely a 2 h rest substantially deteriorating the energy content.

Nanostructural Changes in Zn Anode during Calendar Aging. The large and continuous capacity losses during Zn calendar aging, in both half-cells and full-cells, suggest that the aging mechanism is distinct from that of Li battery chemistries, whose capacity loss during aging is an order of magnitude less and levels off with time. To understand how Zn calendar aging proceeds at the nanoscale, we first employ optical and electron microscopies to examine the current collector substrate in samples with and without 24 h aging in a model electrolyte (2 M $ZnSO_4$). Without calendar aging, the Cu current collector surface appears to be clean and smooth after stripping, as shown by scanning electron microscopy (SEM) and optical microscopy (OM) images in Figure 4a,b. In contrast, dark residues remain on the Cu current collector after stripping with 24 h calendar aging (Figure 4d and Supplementary Fig. 8a). From cross-sectional SEM images, these dark residues appear to be physically disconnected from the current collector substrate (Figure 4d,e and Supplementary Fig. 8b), implying that dead Zn contributes to the ΔCE during 24 h calendar aging. Furthermore, this observed contact loss between Zn and the current collector may be even more severe in other cell geometries, especially for the pouch cell with low external applied pressure.[37]

Selected area electron diffraction (SAED) and lattice imaging in transmission electron microscopy (TEM) confirm that the remaining dark residues after aging are metallic Zn and Zn oxide (Figure 4g–i). As deposited, Zn metal forms in hexagonal morphologies[5,13,40] (Figure 4c). However, local pitting and void formation observed during calendar aging degrade the hexagonal deposits to ill-defined shapes (Figure 4f and Supplementary Fig. 8c,d). This observed degradation suggests the dissolution of Zn^{2+} products from corrosion and implies an additional loss of active material during calendar aging. This is in good agreement with reported X-ray diffraction (XRD) results after calendar aging indicating the formation of $Zn(OH)_2$ and $ZnSO_4[Zn(OH)_2]_3{\cdot}xH_2O$ (ZHS)[26] and also in stark contrast to previous mechanisms of CE loss in Li-based batteries, where the presence of a conformal SEI passivation layer can prevent uninhibited corrosion during calendar aging.[17,19,35,41] Without an SEI layer to effectively passivate Zn metal, corrosion occurs

https://doi.org/10.1021/acsenergylett.3c01282
ACS Energy Lett. 2023, 8, 3820–3828

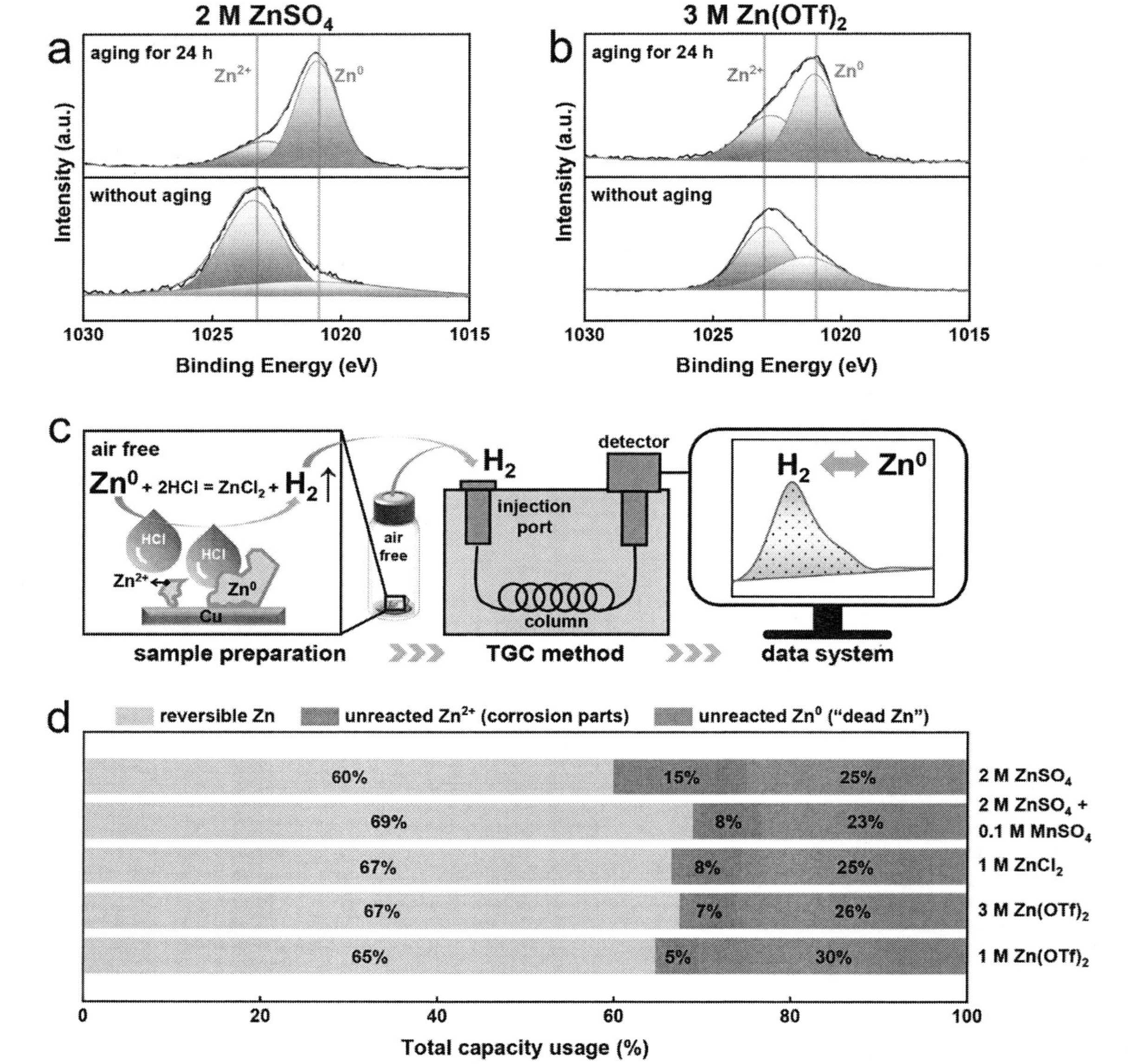

Figure 5. Composition analysis of inactive Zn species left on the Cu foil surface. XPS analysis of samples in (a) 2 M $ZnSO_4$ and (b) 3 M $Zn(OTf)_2$ without and with aging for 24 h. (c) Schematic diagram of the TGC method. (d) Quantitative analysis of capacity usage (reversible Zn, unreacted Zn^{2+} (corrosion parts), and unreacted Zn^0) in various electrolytes by the TGC method, including 2 M $ZnSO_4$, 2 M $ZnSO_4$ with 0.1 M $MnSO_4$ as additive, 1 M $ZnCl_2$, 3 M $Zn(OTf)_2$, and 1 M $Zn(OTf)_2$.

continuously during calendar aging and is coupled to dead Zn formation. The resulting dead Zn and corrosion both contribute to the substantial CE loss, leading to the loss of nearly all deposited capacity after only a week of aging at open circuit voltage (Figure 2d). This is distinct from all previously known mechanisms of calendar aging, which was assumed to proceed primarily through chemical or galvanic corrosion and quantified by coulometry.[15−18,21−25] Here, coulometric and electrochemical methods alone cannot distinguish between corrosion processes and dead Zn formation during calendar aging, making it critical to develop new techniques that can quantify their individual contributions. Such measurements would enable insight into the origins of dead Zn formation and provide possible mitigation strategies.

Differentiating Dead Zn Formation and Corrosion Losses during Calendar Aging. Dead Zn consists of electrically disconnected metallic Zn with an oxidation state of zero. Meanwhile, corrosion products contain Zn components with a +2 oxidation state, such as ZnO, $Zn(OH)_2$, and $ZnSO_4[Zn(OH)_2]_3{\cdot}xH_2O$ (ZHS).[2,20] Zn^0 and Zn^{2+} can be distinguished using X-ray photoelectron spectroscopy (XPS), which we use to examine the Cu current collector surface with and without 24 h aging. In 2 M $ZnSO_4$, both samples exhibit Zn^0 and Zn^{2+} signals of varying strengths located at around

https://doi.org/10.1021/acsenergylett.3c01282
ACS Energy Lett. 2023, 8, 3820−3828

1020.8 eV[8,42,43] and 1023.0 eV,[43−45] respectively (Figure 5a). With calendar aging, the intensity of the Zn^0 peak is much stronger than that of Zn^{2+}, indicating a larger amount of metallic Zn remaining; without calendar aging, this metallic Zn^0 peak is much weaker and suggests that most of the deposited Zn can be stripped reversibly. This result is also observed in other electrolyte chemistries (3 M $Zn(OTf)_2$, 1 M $Zn(TFSI)_2$, and 2 M $ZnSO_4$ + 0.1 M $MnSO_4$), as shown in Figure 5b and Supplementary Fig. 9.

While XPS detects Zn^0 and Zn^{2+} signals with varying intensities, it remains difficult to quantify their contributions to capacity loss with high precision and accuracy. To quantify the contribution of dead Zn and corrosion on the CE loss during calendar aging, we leverage titration gas chromatography (TGC),[44,46−51] which can quantitatively distinguish between any Zn^{2+} and Zn^0 species remaining on the current collector based on the following reaction stoichiometries:

$$Zn + 2HCl \rightarrow ZnCl_2 + H_2\uparrow$$

$$ZnO + 2HCl \rightarrow ZnCl_2 + H_2O$$

$$Zn(OH)_2 + 2HCl \rightarrow ZnCl_2 + 2H_2O$$

As shown in Figure 5c, when we add HCl to the Cu current collector in a sealed container (see method in the Supporting Information), only the reaction between metallic Zn^0 and HCl produces hydrogen gas (H_2). TGC can then quantify the H_2 produced and correlate it directly with the amount of dead Zn remaining on the Cu current collector (Figure 5c). We perform TGC on electrochemically stripped Zn−Cu half-cells after 24 h aging in a diverse range of electrolyte chemistries and reveal that the large majority of CE loss comes from dead Zn (rather than corrosion) in the form of Zn^0, which is consistent with our XPS results. Figure 5d and Supplementary Figs. 10 and 11 summarize our TGC results, quantifying the contributions of dead Zn and corrosion toward capacity losses during calendar aging. The tan-colored bar represents the reversible Zn capacity we can electrochemically strip after 24 h aging, and the red-colored bar represents the TGC-quantified dead Zn remaining on the Cu grid after this stripping process; the blue-colored bar represents the remaining capacity loss due to corrosion, with the sum equaling 100% of the original deposited Zn capacity. Taking two commonly used electrolytes as examples, dead Zn accounts for 25% of the total deposited Zn capacity in 2 M $ZnSO_4$ (40% total CE loss) and 26% in 3 M $Zn(OTf)_2$ (33% total CE loss), which means that the large majority of the CE loss during calendar aging can be attributed to dead Zn. Interestingly, while CE loss due to dead Zn formation remains comparable between the two electrolytes, the CE loss due to corrosion and soluble Zn^{2+} species during calendar aging is less severe in the 3 M $Zn(OTf)_2$ electrolyte (7% CE loss) compared to that in the 2 M $ZnSO_4$ electrolyte (15% CE loss). This suggests corrosion losses can potentially be mitigated by proper electrolyte selection.[9,48] Together, our XPS and TGC results show that dead Zn dominates the CE loss during calendar aging (with corrosion still having a relevant contribution[26]) and elucidates potential design principles to mitigate their impact.

Mitigation Strategies. We find that the absence of a passivating SEI and the formation of electrically disconnected Zn result in large capacity losses from corrosion and dead Zn during the calendar aging. Corrosion can potentially be mitigated by engineering the electrolyte chemistry to expand the voltage window (increasing its stability) or forming a more passivating SEI. Strategies for lowering water's activity (e.g., water-in-salt chemistries or the presence of bulky anions) may be valid here. For example, the 3 M $Zn(OTf)_2$ electrolyte is known to form a more passivating SEI than the 2 M $ZnSO_4$ electrolyte due to the decomposition of the bulky OTf^- anion, which reduces the number of water molecules in the solvation sphere of Zn^{2+} in solution.[33,52] Indeed, the capacity loss due to corrosion for the 3 M $Zn(OTf)_2$ electrolyte (as quantified by TGC) is much lower (~7%) than that of the 2 M $ZnSO_4$ electrolyte (~15%), commensurate with the formation of a more passivating SEI. While quantitative analysis from our TGC experiments reveals that electrolyte chemistries can reduce capacity losses due to corrosion, losses due to dead Zn formation remain relatively constant with varying electrolyte chemistries. Thus, separate approaches that can minimize inactive material formation during calendar aging, such as 3D-structured hosts or high uniaxial pressure to retain electrical connectivity, are potential avenues for exploration. Indeed, we show that a 3D current collector results in lower CE losses (26%) during calendar aging than 2D current collectors (37%) with the same electrolyte chemistry.

Our work presents a detailed and systematic study showing the detrimental impact calendar aging has on the performance of Zn metal batteries. We discover a colossal capacity loss after just 24 h at open-circuit voltage under a variety of conditions (e.g., changing electrolyte chemistry, temperature, aging time, etc.), highlighting the distinct Zn aging mechanisms compared to those of previously known Li-based batteries. In a departure from past studies, we use TGC to untangle the effects of corrosion and dead Zn, quantifying the effect of each process on the measured capacity loss during aging. This allows us to identify important parameters that can potentially mitigate corrosion (e.g., electrolyte chemistry) or dead Zn formation (e.g., current collector architecture). Our results clearly demonstrate that calendar aging poses a key challenge for the practical application of Zn battery chemistries and should not be neglected when evaluating new designs. In addition to cycling performance, the broader implications of our study emphasize the critical importance of quantifying and understanding calendar aging processes in the development of next-generation battery technologies beyond lithium.

■ ASSOCIATED CONTENT

Supporting Information

The Supporting Information is available free of charge at https://pubs.acs.org/doi/10.1021/acsenergylett.3c01282.

Experimental procedures, CE measurements, voltage profiles, TG analysis, XRD, schematic diagrams, cycling performance, SEM, TEM, XPS, calibration curves, and TGC results (PDF)

■ AUTHOR INFORMATION

Corresponding Author

Yuzhang Li − *Department of Chemical and Biomolecular Engineering, University of California, Los Angeles, California 90095, United States; California NanoSystems Institute (CNSI), University of California, Los Angeles, California 90095, United States;* orcid.org/0000-0002-1502-7869; Email: yuzhangli@ucla.edu

Authors

Bo Liu – *Department of Chemical and Biomolecular Engineering, University of California, Los Angeles, California 90095, United States;* orcid.org/0009-0006-1687-656X

Xintong Yuan – *Department of Chemical and Biomolecular Engineering, University of California, Los Angeles, California 90095, United States;* orcid.org/0000-0002-2172-1099

Complete contact information is available at: https://pubs.acs.org/10.1021/acsenergylett.3c01282

Author Contributions

B.L. and Y.L. conceived the ideas. B.L. designed the research under the guidance of Y.L., carried out the experiments, and performed data analysis. X.Y. carried out XPS measurements and helped with SEM and TEM. B.L. and Y.L. wrote the manuscript. All authors discussed the results and commented on the manuscript. All authors have approved the final manuscript.

Notes

The authors declare no competing financial interest.

■ ACKNOWLEDGMENTS

We acknowledge the research support from the Noble Family Innovation Fund and the support of the UCLA-CNSI EICN Facilities. The authors would like to thank the UCLA Molecular Instrumentation Center Mass Spectrometry facility for helping with GCMS instrument operation and method development. We thank Greg Khitrov for his assistance with the TGC experiments design.

■ REFERENCES

(1) Kundu, D.; Adams, B. D.; Duffort, V.; Vajargah, S. H.; Nazar, L. F. A high-capacity and long-life aqueous rechargeable zinc battery using a metal oxide intercalation cathode. *Nat. Energy* **2016**, *1*, 16119.

(2) Pan, H.; Shao, Y.; Yan, P.; Cheng, Y.; Han, K. S.; Nie, Z.; Wang, C.; Yang, J.; Li, X.; Bhattacharya, P.; et al. Reversible aqueous zinc/manganese oxide energy storage from conversion reactions. *Nat. Energy* **2016**, *1*, 16039.

(3) Tang, B.; Shan, L.; Liang, S.; Zhou, J. Issues and opportunities facing aqueous zinc-ion batteries. *Energy Environ. Sci.* **2019**, *12*, 3288–3304.

(4) Wang, S.-B.; Ran, Q.; Yao, R.-Q.; Shi, H.; Wen, Z.; Zhao, M.; Lang, X.-Y.; Jiang, Q. Lamella-nanostructured eutectic zinc-aluminum alloys as reversible and dendrite-free anodes for aqueous rechargeable batteries. *Nat. Commun.* **2020**, *11*, 1634.

(5) Zheng, J.; Zhao, Q.; Tang, T.; Yin, J.; Quilty, C. D.; Renderos, G. D.; Liu, X.; Deng, Y.; Wang, L.; Bock, D. C.; et al. Reversible epitaxial electrodeposition of metals in battery anodes. *Science* **2019**, *366*, 645–648.

(6) Zeng, Y.; Zhang, X.; Qin, R.; Liu, X.; Fang, P.; Zheng, D.; Tong, Y.; Lu, X. Dendrite-Free Zinc Deposition Induced by Multifunctional CNT Frameworks for Stable Flexible Zn-Ion Batteries. *Adv. Mater.* **2019**, *31*, 1903675.

(7) Parker, J. F.; Chervin, C. N.; Pala, I. R.; Machler, M.; Burz, M. F.; Long, J. W.; Rolison, D. R. Rechargeable nickel-3D zinc batteries: An energy-dense, safer alternative to lithium-ion. *Science* **2017**, *356*, 415–418.

(8) Wan, F.; Zhang, L.; Dai, X.; Wang, X.; Niu, Z.; Chen, J. Aqueous rechargeable zinc/sodium vanadate batteries with enhanced performance from simultaneous insertion of dual carriers. *Nat. Commun.* **2018**, *9*, 1656.

(9) Wang, F.; Borodin, O.; Gao, T.; Fan, X.; Sun, W.; Han, F.; Faraone, A.; Dura, J. A.; Xu, K.; Wang, C. Highly reversible zinc metal anode for aqueous batteries. *Nat. Mater.* **2018**, *17*, 543–549.

(10) Cui, Y.; Liu, S.; Liu, B.; Wang, D.; Zhong, Y.; Zhang, X.; Wang, X.; Xia, X.; Gu, C.; Tu, J. Bi-containing Electrolyte Enables Robust and Li Ion Conductive Solid Electrolyte Interphase for Advanced Lithium Metal Anodes. *Front. Chem.* **2020**, *7*, 952.

(11) Kao-ian, W.; Pornprasertsuk, R.; Thamyongkit, P.; Maiyalagan, T.; Kheawhom, S. Rechargeable Zinc-Ion Battery Based on Choline Chloride-Urea Deep Eutectic Solvent. *J. Electrochem. Soc.* **2019**, *166*, A1063.

(12) Konarov, A.; Voronina, N.; Jo, J. H.; Bakenov, Z.; Sun, Y.-K.; Myung, S.-T. Present and Future Perspective on Electrode Materials for Rechargeable Zinc-Ion Batteries. *ACS Energy Lett.* **2018**, *3*, 2620–2640.

(13) Li, Q.; Wang, Y.; Mo, F.; Wang, D.; Liang, G.; Zhao, Y.; Yang, Q.; Huang, Z.; Zhi, C. Calendar Life of Zn Batteries Based on Zn Anode with Zn Powder/Current Collector Structure. *Adv. Energy Mater.* **2021**, *11*, 2003931.

(14) Yuan, X.; Zhang, L.; Li, L.; Dong, H.; Chen, S.; Zhu, W.; Hu, C.; Deng, W.; Zhao, Z.-J.; Gong, J. Ultrathin Pd-Au Shells with Controllable Alloying Degree on Pd Nanocubes toward Carbon Dioxide Reduction. *J. Am. Chem. Soc.* **2019**, *141*, 4791–4794.

(15) Zhang, Z.; Li, Y.; Xu, R.; Zhou, W.; Li, Y.; Oyakhire, S. T.; Wu, Y.; Xu, J.; Wang, H.; Yu, Z.; et al. Capturing the swelling of solid-electrolyte interphase in lithium metal batteries. *Science* **2022**, *375*, 66–70.

(16) Keil, P.; Schuster, S. F.; Wilhelm, J.; Travi, J.; Hauser, A.; Karl, R. C.; Jossen, A. Calendar Aging of Lithium-Ion Batteries. *J. Electrochem. Soc.* **2016**, *163*, A1872.

(17) Boyle, D. T.; Huang, W.; Wang, H.; Li, Y.; Chen, H.; Yu, Z.; Zhang, W.; Bao, Z.; Cui, Y. Corrosion of lithium metal anodes during calendar ageing and its microscopic origins. *Nat. Energy* **2021**, *6*, 487–494.

(18) Wood, S. M.; Fang, C.; Dufek, E. J.; Nagpure, S. C.; Sazhin, S. V.; Liaw, B.; Meng, Y. S. Predicting Calendar Aging in Lithium Metal Secondary Batteries: The Impacts of Solid Electrolyte Interphase Composition and Stability. *Adv. Energy Mater.* **2018**, *8*, 1801427.

(19) Mu, P.; Zhang, H.; Jiang, H.; Dong, T.; Zhang, S.; Wang, C.; Li, J.; Ma, Y.; Dong, S.; Cui, G. Bioinspired Antiaging Binder Additive Addressing the Challenge of Chemical Degradation of Electrolyte at Cathode/Electrolyte Interphase. *J. Am. Chem. Soc.* **2021**, *143*, 18041–18051.

(20) McBrayer, J. D.; Rodrigues, M.-T. F.; Schulze, M. C.; Abraham, D. P.; Apblett, C. A.; Bloom, I.; Carroll, G. M.; Colclasure, A. M.; Fang, C.; Harrison, K. L.; et al. Calendar aging of silicon-containing batteries. *Nat. Energy* **2021**, *6*, 866–872.

(21) Mao, Z.; Farkhondeh, M.; Pritzker, M.; Fowler, M.; Chen, Z. Calendar Aging and Gas Generation in Commercial Graphite/NMC-LMO Lithium-Ion Pouch Cell. *J. Electrochem. Soc.* **2017**, *164*, A3469.

(22) Naumann, M.; Schimpe, M.; Keil, P.; Hesse, H. C.; Jossen, A. Analysis and modeling of calendar aging of a commercial $LiFePO_4$/graphite cell. *J. Energy Storage* **2018**, *17*, 153–169.

(23) Kalaga, K.; Rodrigues, M.-T. F.; Trask, S. E.; Shkrob, I. A.; Abraham, D. P. Calendar-life versus cycle-life aging of lithium-ion cells with silicon-graphite composite electrodes. *Electrochim. Acta* **2018**, *280*, 221–228.

(24) Wu, S.-h.; Lee, P.-H. Storage fading of a commercial 18650 cell comprised with NMC/LMO cathode and graphite anode. *J. Power Sources* **2017**, *349*, 27–36.

(25) Zilberman, I.; Sturm, J.; Jossen, A. Reversible self-discharge and calendar aging of 18650 nickel-rich, silicon-graphite lithium-ion cells. *J. Power Sources* **2019**, *425*, 217–226.

(26) Pu, S. D.; Hu, B.; Li, Z.; Yuan, Y.; Gong, C.; Ning, Z.; Chau, C.; Yang, S.; Zhang, S.; Pi, L.; et al. Decoupling, quantifying, and restoring aging-induced Zn-anode losses in rechargeable aqueous zinc batteries. *Joule* **2023**, *7*, 366–379.

(27) Merrill, L. C.; Rosenberg, S. G.; Jungjohann, K. L.; Harrison, K. L. Uncovering the Relationship between Aging and Cycling on Lithium Metal Battery Self-Discharge. *ACS Appl. Energy Mater.* **2021**, *4*, 7589–7598.

(28) Qiu, H.; Du, X.; Zhao, J.; Wang, Y.; Ju, J.; Chen, Z.; Hu, Z.; Yan, D.; Zhou, X.; Cui, G. Zinc anode-compatible in-situ solid electrolyte interphase via cation solvation modulation. *Nat. Commun.* **2019**, *10*, 5374.

(29) Cui, M.; Fei, J.; Mo, F.; Lei, H.; Huang, Y. Ultra-High-Capacity and Dendrite-Free Zinc-Sulfur Conversion Batteries Based on a Low-Cost Deep Eutectic Solvent. *ACS Appl. Mater. Interfaces* **2021**, *13*, 54981−54989.

(30) Zhu, M.; Hu, J.; Lu, Q.; Dong, H.; Karnaushenko, D. D.; Becker, C.; Karnaushenko, D.; Li, Y.; Tang, H.; Qu, Z.; et al. A Patternable and In Situ Formed Polymeric Zinc Blanket for a Reversible Zinc Anode in a Skin-Mountable Microbattery. *Adv. Mater.* **2021**, *33*, 2007497.

(31) Jia, X.; Liu, C.; Neale, Z. G.; Yang, J.; Cao, G. Active Materials for Aqueous Zinc Ion Batteries: Synthesis, Crystal Structure, Morphology, and Electrochemistry. *Chem. Rev.* **2020**, *120*, 7795−7866.

(32) Zhang, N.; Chen, X.; Yu, M.; Niu, Z.; Cheng, F.; Chen, J. Materials chemistry for rechargeable zinc-ion batteries. *Chem. Soc. Rev.* **2020**, *49*, 4203−4219.

(33) Zhang, N.; Cheng, F.; Liu, Y.; Zhao, Q.; Lei, K.; Chen, C.; Liu, X.; Chen, J. Cation-Deficient Spinel $ZnMn_2O_4$ Cathode in $Zn(CF_3SO_3)_2$ Electrolyte for Rechargeable Aqueous Zn-Ion Battery. *J. Am. Chem. Soc.* **2016**, *138*, 12894−12901.

(34) Liu, B.; Zhang, Y.; Wang, Z.; Ai, C.; Liu, S.; Liu, P.; Zhong, Y.; Lin, S.; Deng, S.; Liu, Q.; et al. Coupling a Sponge Metal Fibers Skeleton with In Situ Surface Engineering to Achieve Advanced Electrodes for Flexible Lithium-Sulfur Batteries. *Adv. Mater.* **2020**, *32*, 2003657.

(35) Lin, D.; Liu, Y.; Cui, Y. Reviving the lithium metal anode for high-energy batteries. *Nat. Nanotechnol.* **2017**, *12*, 194−206.

(36) Zhang, Y.; Deng, S.; Li, Y.; Liu, B.; Pan, G.; Liu, Q.; Wang, X.; Xia, X.; Tu, J. Anchoring MnO_2 on nitrogen-doped porous carbon nanosheets as flexible arrays cathodes for advanced rechargeable Zn-MnO_2 batteries. *Energy Storage Mater.* **2020**, *29*, 52−59.

(37) Ming, F.; Zhu, Y.; Huang, G.; Emwas, A.-H.; Liang, H.; Cui, Y.; Alshareef, H. N. Co-Solvent Electrolyte Engineering for Stable Anode-Free Zinc Metal Batteries. *J. Am. Chem. Soc.* **2022**, *144*, 7160−7170.

(38) An, Y.; Tian, Y.; Zhang, K.; Liu, Y.; Liu, C.; Xiong, S.; Feng, J.; Qian, Y. Stable Aqueous Anode-Free Zinc Batteries Enabled by Interfacial Engineering. *Adv. Funct. Mater.* **2021**, *31*, 2101886.

(39) Zhu, Y.; Cui, Y.; Alshareef, H. N. An Anode-Free Zn-MnO_2 Battery. *Nano Lett.* **2021**, *21*, 1446−1453.

(40) Huang, Z.; Li, Z.; Wang, Y.; Cong, J.; Wu, X.; Song, X.; Ma, Y.; Xiang, H.; Huang, Y. Regulating Zn(002) Deposition toward Long Cycle Life for Zn Metal Batteries. *ACS Energy Lett.* **2023**, *8*, 372−380.

(41) Sato, R. Crystal Growth of Electrodeposited Zinc: An Electron Diffraction and Electron Microscopic Study. *J. Electrochem. Soc.* **1959**, *106*, 206.

(42) Chadwick, D.; Hashemi, T. Adsorbed corrosion inhibitors studied by electron spectroscopy: Benzotriazole on copper and copper alloys. *Corros. Sci.* **1978**, *18*, 39−51.

(43) Diler, E.; Lescop, B.; Rioual, S.; Nguyen Vien, G.; Thierry, D.; Rouvellou, B. Initial formation of corrosion products on pure zinc and $MgZn_2$ examinated by XPS. *Corros. Sci.* **2014**, *79*, 83−88.

(44) Zou, P.; Lin, R.; Pollard, T. P.; Yao, L.; Hu, E.; Zhang, R.; He, Y.; Wang, C.; West, W. C.; Ma, L.; et al. Localized Hydrophobicity in Aqueous Zinc Electrolytes Improves Zinc Metal Reversibility. *Nano Lett.* **2022**, *22*, 7535−7544.

(45) Strohmeier, B. R.; Hercules, D. M. Surface spectroscopic characterization of the interaction between zinc ions and γ-alumina. *J. Catal.* **1984**, *86*, 266−279.

(46) Hobold, G. M.; Gallant, B. M. Quantifying Capacity Loss Mechanisms of Li Metal Anodes beyond Inactive Li0. *ACS Energy Lett.* **2022**, *7*, 3458−3466.

(47) Paul, P. P.; McShane, E. J.; Colclasure, A. M.; Balsara, N.; Brown, D. E.; Cao, C.; Chen, B.-R.; Chinnam, P. R.; Cui, Y.; Dufek, E. J.; et al. A Review of Existing and Emerging Methods for Lithium Detection and Characterization in Li-Ion and Li-Metal Batteries. *Adv. Energy Mater.* **2021**, *11*, 2100372.

(48) Sreenarayanan, B.; Tan, D. H. S.; Bai, S.; Li, W.; Bao, W.; Meng, Y. S. Quantification of lithium inventory loss in micro silicon anode via titration-gas chromatography. *J. Power Sources* **2022**, *531*, 231327.

(49) Gao, N.; Abboud, A. W.; Mattei, G. S.; Li, Z.; Corrao, A. A.; Fang, C.; Liaw, B.; Meng, Y. S.; Khalifah, P. G.; Dufek, E. J.; et al. Fast Diagnosis of Failure Mechanisms and Lifetime Prediction of Li Metal Batteries. *Small Methods* **2021**, *5*, 2000807.

(50) Fang, C.; Lu, B.; Pawar, G.; Zhang, M.; Cheng, D.; Chen, S.; Ceja, M.; Doux, J.-M.; Musrock, H.; Cai, M.; et al. Pressure-tailored lithium deposition and dissolution in lithium metal batteries. *Nat. Energy* **2021**, *6*, 987−994.

(51) Lu, B.; Olivera, E.; Scharf, J.; Chouchane, M.; Fang, C.; Ceja, M.; Pangilinan, L. E.; Zheng, S.; Dawson, A.; Cheng, D.; et al. Quantitatively Designing Porous Copper Current Collectors for Lithium Metal Anodes. *ACS Appl. Energy Mater.* **2021**, *4*, 6454−6465.

(52) Li, D.; Cao, L.; Deng, T.; Liu, S.; Wang, C. Design of a Solid Electrolyte Interphase for Aqueous Zn Batteries. *Angew. Chem., Int. Ed.* **2021**, *60*, 13035−13041.

444d Hydrated Eutectic Electrolytes for High Performance Rechargeable Zinc Batteries

Raju Vadthya, Olumide Oladoyin, Hao Nguyen, Shuya Wei

Zinc has attracted significant interest as a promising anode material in aqueous rechargeable batteries, offering advantages such as high volumetric capacity (5857 $mAhcm^{-3}$ vs 2042 $mAhcm^{-3}$ for Li), low cost (2.4 $USD.kg^{-1}$ vs. 19.2 $USD.kg^{-1}$ for Li), and multiple electrons transfer capability ($2e^-$ vs. $1e^-$ for Li). However, the intractable side reactions at the negative Zn anode and parasitic hydrogen evolution in the aqueous media hamper the technology unattainable for practical evaluations. Here, we present a hydrated eutectic electrolyte by combining hydrated Zn salt ($ZnSO_4.7H_2O$) with triethylamine hydrochloride. Eutectic formation between the two salts is attributed to Lewis acid-base interactions, facilitated by substantial charge delocalization exhibited by the SO_4^- anions, serving as a strong Lewis bases. The bipolar nature of H_2O readily interacts with the Zn^{+2}, forming an ion-conducting coordination complex, thereby suppressing the hydrogen evolution reaction (HER). The electrolyte offers a high electrochemical stability window (2.45V vs. Zn/Zn^{+2}), wide liquid range suitable (T_{liquid} ^{3}-20 °C), low vapor-pressure, environment insensitivity and the symmetrical Zn||Zn cell demonstrates excellent stability for >1000 hours at 0.1 mA/cm^2. When paired with V_2O_5, Zn exhibits a high discharge capacity of 270 mAh/g, and promising cyclability for >600 cycles. These results highlight its potential as an alternative to aqueous electrolytes for grid-level storage units.

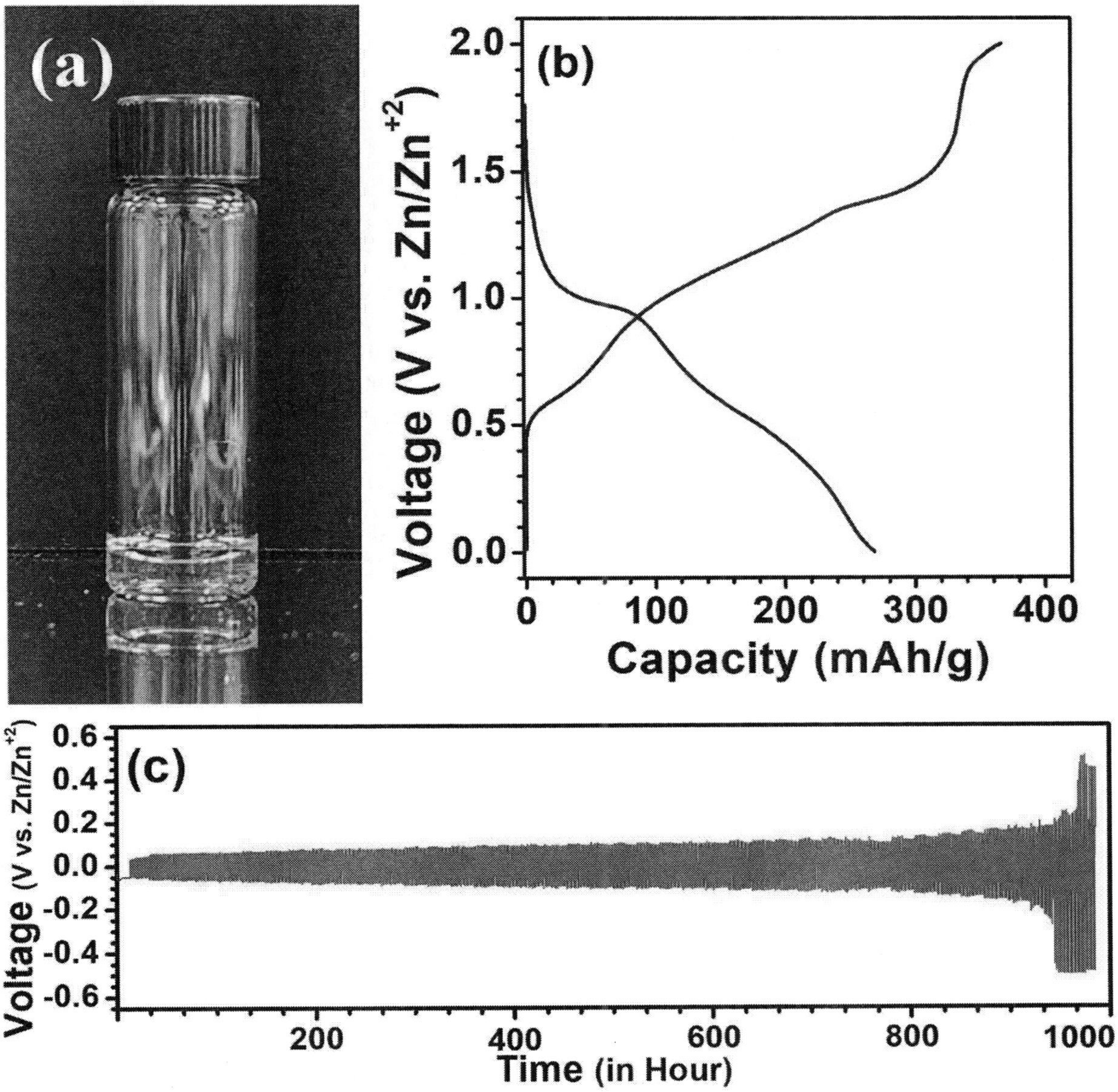
(a)
(b)
Voltage (V vs. Zn/Zn^{+2})
Capacity (mAh/g)
(c)
Voltage (V vs. Zn/Zn^{+2})
Time (in Hour)

444e Methacrylate Based Ionically Conductive Polymer Electrolyte for Solid State Lithium Iron Phosphate Batteries

Wissam Fawaz, Alexander Hoffmann, Simon Ng

Lithium Iron Phosphate (**LFP**) is a promising electrode material for energy storage that is widely viewed as a thermally resistant alternative to current state of the art lithium-ion cathode materials employed in modern electrification efforts. Solid state LFP batteries offer desirable safety, lacking from many alternatives, as it nullifies the use of flammable organic electrolytes. To date, solid-state electrolytes developed for LFP cells have been lacking in room temperature ionic conductivity, largely due to high interfacial impedance and reactivity of the solid electrolyte with electrode surfaces. However, methyl methacrylates (**MMA**) are the nascent trend in the realm of solid electrolyte membranes because of their low cost, malleability and reported stability with lithium metal interfaces. Through the integration of lithium salt and fillers into the conductive framework of an MMA polymer, a flexible UV curable gel polymer electrolyte was formed with promising ionic conductivity characterized through electrochemical impedance spectroscopy (EIS). The crystallinity and morphology of the formed electrolyte was investigated via X-ray diffraction (XRD) and Scanning Electron Microscopy (SEM) respectively. Electrochemical performance of the electrolyte was demonstrated in a system comprising of an LFP electrode paired with a lithium metal anode that offers a stable, room temperature, cycling capacity of 80 mAh/g. This study will provide a better understanding of the interaction of lithium salt and fillers with MMA electrolytes on the energy density and cycling stability in solid state lithium-ion batteries.

444f 3D Printed Nanoporous Separators for High Cycling Stability Li-Ion Batteries

Jiandi Wan

Separators with controllable microstructures are a crucial component of high-power high-energy Li-ion batteries. Microstructural inhomogeneities within the separator have been linked to battery failure by localized current hot spots. Here, we demonstrate the synergetic effect of using projection micro-stereolithography and polymerization induced phase separation to fabricate hexanediol diacrylate (HDDA) separators with controlled multi-length-scale morphologies and high porosities (26.6% - 63.1%). Systematic variation of the microstructure of the HDDA separators enabled the examination of structure-property relationships with respect to battery stability and performance. The graphite/NMC 622 cell assembly with the HDDA separator showed better high-rate capability and capacity retention than the Celgard 2325 separator, delivering a reversible capacity of 143 mAh/g with a 99.85% capacity retention after 100 cycles at 1C.

444h High-Capacity Micron Si/Graphite/Vanadium Carbide Composite As Anode for Li-Ion Batteries

Rohit Choudhury, Praveen Meduri

The ever-increasing demand for portable consumer electronics and renewable energy storage devices has spurred global interest in high-performance lithium-ion batteries (LIBs). The development of LIBs with enhanced capacity, extended cycle stability, and reduced weight forms the critical strategies necessary to cater to the requirements of next-generation devices [1]. Despite graphite being the commercial anode in LIBs, its relatively low theoretical capacity (372 mAh g^{-1}) limits its utilization in high-energy applications. Silicon (Si), with its high theoretical capacity of 4200 mAh g^{-1}, natural abundance, and suitable Li uptake potential (0.4 V vs Li/Li+), emerges as a promising anode material [2]. However, its extensive volume expansion hinders its practical application.

In this work, we combine micron-sized Si particles with graphite. The use of micron Si is cost-effective, and the synthesis process is straightforward. The finely ground mixture (Si-G) is combined with polyacrylonitrile (PAN) to yield amorphous carbon upon carbonization. This assists in buffering the mechanical stresses induced by Si volume expansion. Moreover, it prevents the direct exposure of Si to the electrolyte, thereby stabilizing the solid electrolyte interphase (SEI). The incorporation of a dopant enhances the material's intrinsic conductivity, leading to an increased capacity by facilitating electron and Li-ion transport channels. To further boost the composite's performance, synthesized Mxene is incorporated to augment the charge transport between micron Si. The composite's morphology and structure are examined using various techniques such as SEM-EDS, TEM, XRD, Raman, and XPS spectroscopy, revealing a uniform distribution of Si, C, and Mxene sheets that maintain the sheet architecture.

The composite, when used as an anode, exhibits exceptional electrochemical performance, as evidenced by its remarkable lithium storage specific capacity of 2003 mAh g^{-1} (based on the weight of Si) even after prolonged cycling (at 1C rate). At the end of 500 cycles, the volume expansion was reduced to only about 150% of its initial size (from 420%), with an almost intact morphology. Additionally, the composite demonstrates superior rate performance with a specific capacity of ~2440 mAh g^{-1} at a 10 C rate, highlighting its potential for high-power applications. Furthermore, the electrode exhibited a low charge transfer impedance and rapid electron transport, which enhanced the electrochemical performance. The improved performance can be attributed to the unique structure comprising Mxene and doping, which significantly augmented the Li-ion diffusion and the active sites. Also, the use of low-cost micron Si with a high tap density is employed. Consequently, this study offers a methodology to develop high-capacity LIBs based on micron silicon.

References

1. Liu, L. Kang, J. Hu, E. Jung, J. Zhang, S.C. Jun, Y. Yamauchi, ACS Energy Lett. 6 (2021) 3011–3019.
2. Zhou, Y. Liu, C. Du, Y. Ren, T. Mu, P. Zuo, G. Yin, Y. Ma, X. Cheng, Y. Gao, Journal of Power Sources. 381 (2018) 156–163.

733a Interfaces and Interphases in Next-Generation Battery Technologies

Jeffrey Lopez

While lithium based battery technologies are becoming increasingly widespread in our energy landscape, both in electric vehicles and grid scale storage, room for improvement remains in continuing to increase energy density, reducing their costs, and improving the sustainability of their manufacturing. In this presentation, we will discuss recent progress in our group toward achieving these goals. Specifically, efforts to comprehensively characterize the reaction intermediates and organic SEI products of electrolyte reduction reactions *in operando* and *ex situ* will be detailed. Both EPR and ATR-FTIR spectroscopy are used to identify electrolyte reaction intermediates and products to clarify electrolyte reduction mechanisms for Li metal batteries. A particular focus is placed on fluorinated electrolytes, which have shown high performance in Li metal batteries but pose environmental concerns related to the longevity of PFAS compounds. Improved understanding of electrolyte reduction mechanisms will help to chart a path toward fluorine free electrolytes. We will also discuss the study of electrode-binder interfaces in critical-mineral free conversion cathodes and share findings of targeted binder-active material interactions that improve both reversibility and cycling stability of these promising electrode materials. With the understanding developed from studying these critical interfaces and interphases in promising electrode materials, we provide new insights into the formation and chemical nature of interphase components that promote stable cycling of high energy density lithium metal and conversion electrodes. Together, these chemistries offer a pathway toward high energy density and low cost storage while avoiding the use of critical minerals such as Ni and Co.

733b Growth Front Evolution in Electrodeposition on 3D Frameworks Fabricated Via Dhbt

Itishree Panda, Trina Dhara, Sunando DasGupta

To cope up with the surge in energy demand, metal anode batteries are increasingly replacing the conventional metal ion batteries due to their higher specific energy density. During electrochemical charging of solid-state batteries, metal ions (Ca, Mg, Na, K, Li) get deposited on the current collector as pure metallic anode. The plating of the anode experiences a dynamic nonplanar metal front evolution as a result of non-uniform current distribution, namely dendrites. The metal anode undergoes expansion/shrinkage during charging/discharging, eventually structurally hampering the battery. Moreover, the durability and safety of the batteries are compromised when the growth front of the anode ultimately penetrates the separator and reaches the cathode, causing short-circuiting of the cell and possible battery thermal runaway. Although the use of different electrolytes, electrolyte additives, and charging protocols etc. have proven to be beneficial in combating the anode deformation and undesirable metallic growth up to some extent, a favorable design in 3D current collectors situated below the metal anode will further alleviate the issues. Herein, morphology dependent electrodeposition of copper on structured frameworks is explored that may provide insights in designing the current collectors. Dynamic hydrogen bubble template (DHBT) method is utilized to fabricate rough, honeycomb structured and porous micro/nano structured copper electrodes. In-situ 2D visualization of the dynamic evolution front (metallic copper) exhibits distinct morphologies (Mossy like, closely packed structures) owing to the structure mediated local surface current density distribution. Porous electrodes with high internal surface area show a reduction in deposition height of the metallic front. The cumulative effect of minimized local effective current density and homogenous distribution of incoming metal atoms over the entire area manifests itself in the reduction of unwanted metallic growth.

733c A New Charging Algorithm for Lithium-Ion Batteries Based on a Constant Lithium Gradient Protocol

Kasim Adewuyi

This paper presents a new optimal charging methodology called Constant Gradient Constant Voltage (CGCV), which limits mechanical crack propagation at the surface of the particles in electrodes by regulating the current to maintain a constant lithium gradient at the surface. One important cause of battery degradation is the stress developed by the concentration gradient inside active particles. The proposed new charging protocol actively controls the concentration gradient by using the lithium diffusivity as strongly state-of-charge (SOC) dependent. Here, the diffusivity is not real time measurable, so a modeling-based approach is incorporated, in which the charging protocol is guided based on the concentration gradient predicted from the modeling results. The upper limit of the concentration gradient is defined based on the maximum stress during the conventional Constant Current Constant Voltage (CCCV) charging. Two gradient options of Mid and High (which is set by the maximum stress limit relative to 0.1C). The experimental results under three charging scenarios of high CGCV, mid CGCV, and standard CCCV. CGCV charging methodologies show slightly slower degradation rates with a 56% (High CG) or 49% (Mid CG) reduction in charging time when comparing CC to CG. This is a new protocol considering battery physics on particle level and has great potential to address the key challenge of regulating battery degradation during fast charging of Electric Vehicles (EVs).

Generally, charging methodologies are experimental approaches with the primary focus on improving cycle life and do not directly consider specific degradation mechanisms. There is no effort to improve mechanical degradation, which is a critical degradation mechanism. The extraction and insertion of ions distorts the structure of the electrodes. Analysis via TEM has shown that if there is insufficient time for the ions to diffuse through the electrode, inhomogeneity results in stress being relieved via fracturing. Research into $LiCoO_2$ shows that a model can determine a critical fracture point based on particle size, diffusivity, and current rate by solving the elastic boundary conditions . Therefore, the main method to prevent crack propagation is to lower the critical facture energy by reducing the particle size or the C-rate. This model determines critical particle size is 250 nm at 2C and decreases to 190 nm at 6C during discharge. Another important factor to consider is if any phase shift occurs for the electrode particles. Another stress fracture model studying $LiMn_2O_4$ shows that if Jahn-teller distortion occurs during the transition from spinel to cubic then fracture is unavoidable. This occurs due to the mismatch in lattice parameters between the two crystal structures causing an immediate stress high enough to induce fracture. This region can be avoided for particles between 500 nm to 800 nm during charging if the cell is kept above 3 V. However, if the particle is kept in single phase, then fracture tends to occur in 4 V plateau region if current exceeds 5.36C for 5 nm particles . Though lesser subcritical fracturing can occur due to surface interactions between cracks and the electrolyte due to natural wear and tear.

This work proposes an approach designed to address that issue. An electrode is assumed to have preexisting cracks on its surface; these cracks will not grow when the maximum energy release is smaller than the fracture energy. This maximum energy release is caused by the gradient of Li ion distribution determined by the relationship between diffusivity and flux. Here, the diffusivity is strongly dependent on SOC, so the proposed idea is to regulate the gradient by varying the current based on the SOC dependent diffusivity. Specifically, the concentration gradient is controlled and is fixed to a certain value based on a CC profile. This is called Constant Gradient (CG), which is designed to allow higher current input during the high diffusivity SOC regions, while minimizing the stress development in the low diffusivity regions by using a smaller current input. For ideal implementation, the SOC dependent variable diffusivity would be accessible on-line for the control, but this is difficult to directly measure especially when the cell is running. Therefore, a modeling-assisted approach is incorporated instead to provide the current profile by considering the SOC-dependent diffusivity.

733e Non-Equilibrium Continuum Modeling of Energy Recovery in Forward-Biased Bipolar Membranes

Justin Bui, Eric Lees, Alexis T. Bell, Adam Weber

The ability for bipolar membranes (BPMs) to interconvert voltage and pH makes them attractive materials for use in energy conversion and storage. Reverse-biased BPMs, which use electrical voltage to dissociate water into acid and base, have become increasingly well-studied for applications where sustainable acid and base generation is crucial. However, forward-biased BPMs (FB-BPMs), in which electrical work is extracted from pH gradients through H^+/OH^- recombination, are still poorly understood. This is unfortunate, because FB-BPMs attractive for a variety of applications. In CO_2 reduction, FB-BPMs furnish optimal alkaline cathode environments for CO_2 electrolysis, mitigate the crossover of (bi)carbonate species, and enable recovery of energy from acid-base recombination to reduce energy requirements for electrolysis. FB-BPMs also overcome challenges with membrane dehydration in hydrogen fuel cells by generating water at the CL to continuously humidify the cell. Finally, FB-BPMs can be used increase the energy density of redox flow batteries (RFBs) by providing an additional source of voltage (beyond the faradaic reactions at the electrodes), enabling operation outside the water stability region. Enabling these applications requires significantly higher current and power densities for FB-BPMs than the current state-of-the-art. Hence, an improved understanding of the physical phenomena that limit current density and energy recovery in these materials is needed.

In this work, an experimentally-validated continuum modeling framework is developed to elucidate mechanisms of transport and recombination in FB-BPMs. Simulations reveal that open-circuit voltage (OCV) in FB-BPMs is a dynamic property controlled by the intricate balance of ion-recombination and crossover, and that the presence of competitive counter-ion impurities significantly attenuates limiting current density of FB-BPMs through selective competition with H^+ and OH^- for sites within the ion-conducting polymer layers. Additionally, fixed-charge neutralization reduces the concentration of available sites that mediate the transport of H^+ and OH^-, further diminishing limiting current densities. Collectively, the established theory underscores the importance of non-equilibrium transport and kinetic phenomena in dictating the efficiency and performance of FB-BPMs for energy recovery. The implications of the simulations are discussed in the context of CO_2 electrolysis systems and energy storage systems that employ FB-BPMs, and guidelines for the development of future FB-BPM materials and devices are presented.

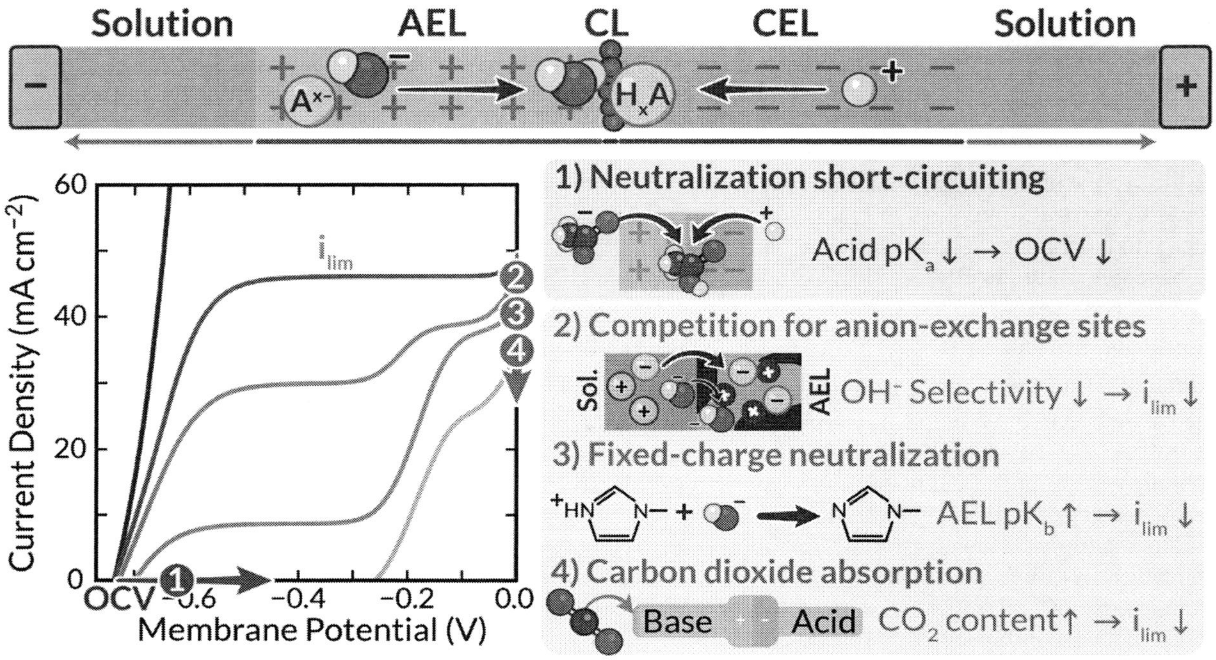
Solution
AEL
CL
CEL
Solution
A^{x-}
H_xA
Current Density (mA cm^{-2})
Membrane Potential (V)
i_{lim}
OCV
1) Neutralization short-circuiting
Acid pK_a ↓ → OCV ↓
2) Competition for anion-exchange sites
Sol.
AEL
OH^- Selectivity ↓ → i_{lim} ↓
3) Fixed-charge neutralization
AEL pK_b ↑ → i_{lim} ↓
4) Carbon dioxide absorption
Base
Acid
CO_2 content ↑ → i_{lim} ↓

733f Solar-Driven Thermochemical Utilization of CO_2 to Produce Fuels Using Ni-Doped Iron Oxides: Experimental Evaluation and Efficiency Analysis

Rahul Bhosale, Sayma Akhter

The Fischer-Tropsch process is an established method used to produce liquid transportation fuels, including gasoline, diesel, and jet fuel. This process involves the conversion of syngas, which is a mixture of hydrogen and carbon monoxide, into liquid hydrocarbons. Syngas is typically produced by gasification of coal, natural gas, or biomass. The excessive utilization of fossil fuels has become a major concern in recent times. Apart from depleting their finite resources, the widespread use of fossil fuels has also led to a significant increase in environmental issues. These issues include air pollution, water pollution, and the release of greenhouse gases that contribute to global warming and climate change. The negative impacts of these environmental issues can be felt on a global scale, affecting our health, ecosystems, and the overall sustainability of our planet. As such, it's crucial that we explore alternative sources of energy and adopt more sustainable practices to minimize our reliance on fossil fuels and mitigate their harmful effects on the environment. An option to create syngas is to use redox reactions based on doped iron oxide, which involves splitting H_2O and CO_2. The process consists of two steps. The first step involves the thermal reduction of iron oxides that have been doped. In the second step, the doped iron oxides are re-oxidized through the splitting of H_2O and/or CO_2, which results in the production of fuel such as H_2 or syngas. In this study, the production of Ni-doped iron oxides was achieved through a combustion synthesis method. The synthesized Ni-doped iron oxides were subjected to characterization and testing in a high-temperature thermogravimetric analyzer. The study examined the effect of operating parameters on long-term redox reactivity and fuel production capacity. Additionally, a computational thermodynamic analysis was conducted to determine the solar-to-fuel energy conversion efficiency of the Ni-doped iron oxide-driven solar thermochemical CO_2 conversion cycle.

AUTHOR INDEX